ETABS 2016 Black Book

By
Gaurav Verma
(CADCAMCAE Works)

Edited by
Kristen

ISBN # 978-1-988722-29-0

NOTICE TO THE READER

DEDICATION

To teachers, who make it possible to disseminate knowledge
to enlighten the young and curious minds
of our future generations

To students, who are the future of the world

THANKS

To my friends and colleagues

To my family for their love and support

Table of Contents

Preface xi
About Author xiii

Chapter 1 : Starting with ETABS

Introduction — 1-2
Downloading and Installing ETABS 2016 Evaluation — 1-2
Starting ETABS — 1-4
Starting a New Model — 1-5
Grid Dimensions — 1-6
Story Dimensions — 1-7
Add Structural Objects — 1-8
User Interface of ETABS — 1-12
Menu bar — 1-12
Toolbar — 1-12
Status Bar — 1-12
Working Plane drop-down — 1-12
Story drop-down — 1-12
Grid System drop-down — 1-13
Units — 1-13
Model Explorer — 1-13
Model Tab — 1-13
Display Tab — 1-18
Tables Tab — 1-19
Reports Tab — 1-20
Detailing Tabs — 1-21
File Menu — 1-21
Creating a New Model — 1-21
Opening a Model File — 1-21
Closing a Model File — 1-22
Saving a File — 1-22
Importing Files/Data — 1-22
Exporting Files — 1-27
Upload to CSI Cloud — 1-28
Creating Video of Analysis — 1-28
Printing Graphics — 1-29
Generating Reports — 1-32
Capturing Picture — 1-35
Checking the Project Information — 1-36

Chapter 2 : Starting a Project in ETABS

Introduction — 2-2
Workflow in Etabs — 2-2
Modifying Grid — 2-2
Setting Grid Options — 2-2
Editing Stories and Grid System — 2-3
Drawing Joint Objects — 2-3
Adding Grid Lines at Selected Joints — 2-4

Drawing Grid Lines Manually   **2-5**
Snap Options   **2-6**
Modifying Snap Options   2-6
Enabling/Disabling Snapping Function   2-7
Drawing with Snap Only   2-7
Drawing Beams/Columns/Brace Objects   **2-7**
Drawing Beams/Columns/Braces in Plan, Elev, and 3D   2-8
Quick Draw Beams/Columns (Plan, Elev, 3D)   2-10
Quick Draw Secondary Beams (Plan, 3D)   2-10
Quick Draw Braces (Plan, Elev, 3D)   2-11
Drawing Floor and Wall Objects   **2-11**
Draw Floor/Wall (Plan, Elev, 3D) Tool   2-12
Draw Rectangular Floor/Wall (Plan, Elev) Tool   2-12
Quick Draw Floor/Wall (Plan, Elev) Tool   2-13
Draw Walls (Plan) Tool   2-14
Drawing Links   **2-15**
Drawing Tendons   2-15
Drawing Design Strips   2-15
Drawing Dimension Lines   **2-16**
Drawing Reference Points   **2-17**
Setting Views in Viewports   **2-17**
Setting 3D View   2-17
Setting Plan View   2-17
Setting Elevation View   2-18
Drawing Reference Planes   **2-19**
Draw Developed Elevation Definition   **2-19**
Drawing Wall Stack   **2-21**
Auto Draw Cladding   **2-22**

Chapter 3 : Section Properties and Material

Introduction   **3-2**
Defining Section Properties   **3-2**
Defining Frame Sections   3-2
Section Designer   **3-11**
Activating Snaps   3-12
Creating Reference Geometries   3-13
Drawing Shapes   3-14
Checking Properties of Section   3-18
Defining Tendon Sections   **3-20**
Adding a New Tendon Property   3-21
Adding Copy of a Tendon Property   3-21
Modifying a Tendon Property   3-22
Deleting a Tendon Property   3-22
Defining Slab Sections   **3-22**
Defining Deck Sections   **3-25**
Defining Wall Section   **3-26**
Defining Reinforcing Bar Sizes   **3-26**
Defining Link/Support Properties   **3-27**
Defining Frame/Wall Hinge Properties   **3-29**
Defining Panel Zone   **3-31**

Defining Springs Properties **3-33**
Defining Point Spring Properties 3-33
Defining Line Spring and Area Spring Properties 3-34
Soil Profile 3-34
Properties of Isolated Column Footing 3-36
Diaphragms Properties **3-37**
Pier Labels and Spandrel Labels **3-37**
Creating Groups **3-38**

Chapter 4 : Assigning Properties and Applying Loads

Introduction 4-2
Assigning Properties to Joints **4-2**
Assigning Restraint Joint Properties 4-2
Assigning Spring Joint Properties 4-3
Assigning Diaphragms Joint Properties 4-4
Assigning Panel Zone Properties to Joints 4-5
Assigning Additional Mass to Selected Joints 4-5
Setting Joint Floor Meshing Options 4-6
Assigning Frame Properties **4-7**
Assigning Section Properties 4-7
Assigning Property Modifier to Frame 4-8
Releasing or Partially Fixing Ends of Frame Members 4-9
Assigning End Length Offset 4-10
Assigning Insertion Points for Frame Members 4-11
Assigning Local Axes 4-12
Assigning Output Stations 4-13
Assigning Tension/Compression Limits 4-14
Assigning Hinges 4-15
Assigning Hinge Overwrites 4-16
Assigning Line Spring 4-17
Assigning Additional Mass to Frame Members 4-17
Assigning Pier Labels 4-18
Frame Auto Mesh Options 4-18
Defining Moment Connections of Frame Beams 4-19
Assigning Column Splice Overwrites 4-20
Assigning Nonprismatic Property Parameters 4-20
Assigning Material Overwrites 4-21
Assigning Column/Brace Rebar Ratio for Creep Analysis 4-22
Assigning Shell Properties **4-22**
Assigning Slab Section 4-23
Assigning Deck Section Properties 4-24
Assigning Wall Section 4-24
Assigning Openings 4-25
Assigning Stiffness Modifiers 4-25
Assigning Thickness Overwrites 4-26
Assigning Insertion Point to Shell Objects 4-27
Assigning Diaphragms to Shell Objects 4-28
Assigning Edge Releases 4-29
Assigning Local Axes 4-30

Assigning Area Springs   4-30
Assigning Additional Mass to Shell Objects   4-31
Assigning Link Properties   **4-31**
Assigning Link Properties   4-32
Orienting Local Axes for the Links   4-32
Assigning Properties to Tendons   **4-33**
Assigning Joint Loads   **4-33**
Applying Force at the Joint   4-34
Applying Ground Displacement at the Joints   4-35
Assigning Temperature to Joints   4-35
Assigning Frame Loads   **4-36**
Assigning Point load on Frame Members   4-36
Assigning Distributed Load on Frame Members   4-37
Assigning Temperature to Selected Frame Members   4-38
Assigning Open Structure Wind to Frame Members   4-38
Assigning Shell Loads   **4-39**
Assigning Uniform Load Sets   4-39
Assigning Uniform Shell Load   4-41
Assigning Non-uniform Load to Shell Objects   4-41
Applying Temperature to Shell Objects   4-42
Assigning Loads to Tendons   **4-42**
Assigning Tendon Loads   4-43
Assigning Tendon Losses   4-43
Copying and Pasting Assigns   **4-44**

Chapter 5 : Creating Load Cases and Performing Analysis

Introduction   **5-2**
Defining Load Patterns   **5-2**
Modifying Lateral load of wind   5-3
Creating Modal Cases   **5-6**
Setting Load Cases   **5-8**
Creating Linear Static Load Case   5-9
Creating Nonlinear Static Load Case   5-11
Creating Nonlinear Staged Construction Load Case   5-14
Creating Response Spectrum Load Case   5-16
Creating Time History Load Case   5-17
Creating Buckling Load Case   5-22
Creating Hyperstatic Load Case   5-23
Setting Load Combinations   **5-24**
Adding New Combination   5-24
Adding Default Design Combinations in the List   5-25
Converting Combination to Nonlinear Case   5-26
Creating Auto Construction   **5-26**
Sequence Load Case   **5-26**
Creating Walking Vibrations   **5-28**
Checking Model for Analysis   **5-29**
Setting Degree of Freedom for Analysis   **5-30**
Setting Load Cases to Run Analysis   **5-31**
Setting SAPFire Solver Options   **5-32**
Automatic Mesh Settings for Floors   **5-33**

Setting Mesh Size for Walls   5-34
Setting Hinges for Analysis Model   5-34
Cracking Analysis Options   5-35
Running the Analysis   5-36
Applying Model Alive   5-36
Modifying Undeformed Geometry   5-37
Checking Analysis Log   5-38
Unlocking model   5-38
Display Options   5-38
Undeformed Shape   5-38
Load Assigns   5-39
Deformed Shape   5-39
Force/Stress Diagrams   5-39
Display Performance Check   5-43
Energy/Virtual Work Diagram   5-43
Cumulative Energy Components   5-44
Story Response Plots   5-45
Plot Functions   5-45

Chapter 6 : Running Design Study

Introduction   6-2
Steel Frame Design   6-2
Steel Frame Preferences   6-2
View/Revise Overwrite for Steel Frame Sections   6-5
Setting Lateral Brace Parameters   6-5
Selecting Design Group   6-7
Selecting Design Combinations   6-7
Performing Design Check   6-8
Stress Information of Structural members   6-9
Interactive Design View   6-11
Display Design Info   6-11
Removing Auto Select Sections   6-12
Changing Steel Frame Sections   6-12
Resetting the Design Sections to Previous Analysis Values   6-13
Verifying Analysis V/s Design Section   6-13
Verifying All Members according to Analysis Parameters   6-14
Resetting All Overwrites   6-14
Deleting Design Results   6-14
Concrete Frame Design   6-14
Composite Beam Design, Composite Column Design, and
Steel Joist Design   6-15
Overwrite Frame Design Procedure   6-15
Shear Wall Design and Concrete Slab Design   6-17
Live Load Reduction Factors   6-18
Setting Lateral Displacement Targets   6-20
Setting Time Period Targets   6-20

Chapter 7 : Detailing

Introduction   7-2
Detailing Preferences   7-2

Concrete Detailing Preferences	7-3
Steel Detailing Preferences	7-3
Setting Rebar Selection Rule	7-4
Setting Rebar Rules for Beams	7-4
Adding/Modifying Slab Sections	7-5
Drawing Section Line	7-5
Deleting Selected Sections	7-6
Show/Modify Section Line Properties	7-7
Drawing Sheet Setup	7-7
Starting Detailing	7-8
Show Detailing	7-9
Clear Detailing	7-10
Exporting Drawing	7-10
Printing Drawings	7-11

Chapter 8 : Project

Project	8-2

Preface

ETABS is the ultimate integrated software package for the structural analysis and design of buildings. Incorporating 40 years of continuous research and development, this latest ETABS offers unmatched 3D object based modeling and visualization tools, blazingly fast linear and nonlinear analytical power, sophisticated and comprehensive design capabilities for a wide-range of materials, and insightful graphic displays, reports, and schematic drawings that allow users to quickly and easily decipher and understand analysis and design results.

The **ETABS 2016 Black Book,** is written to help beginners learn the basics of ETABS structure modeling and analysis. The book follows a step by step methodology. This book explains the designing of structure, assigning various properties to structure, applying different load conditions, and performing analyses. This book also covers the basics of detailing in ETABS. The book covers almost all the information required by a learner to master ETABS. Some of the salient features of this book are:

In-Depth explanation of concepts

Every new topic of this book starts with the explanation of the basic concepts. In this way, the user becomes capable of relating the things with real world.

Topics Covered

Every chapter starts with a list of topics being covered in that chapter. In this way, the user can easy find the topic of his/her interest easily.

Instruction through illustration

The instructions to perform any action are provided by maximum number of illustrations so that the user can perform the actions discussed in the book easily and effectively. There are about 400 illustrations that make the learning process effective.

Tutorial point of view

The book explains the concepts through the tutorial to make the understanding of users firm and long lasting. Each chapter of the book has tutorials that are real world projects.

Project

Free projects and exercises are provided to students for practicing.

For Faculty

If you are a faculty member, then you can ask for video tutorials on any of the topic, exercise, tutorial, or concept.

Formatting Conventions Used in the Text

All the key terms like name of button, tool, drop-down etc. are kept bold.

Free Resources

Link to the resources used in this book are provided to the users via email. To get the resources, mail us at *cadcamcaeworks@gmail.com* or *info@cadcamcaeworks.com* with your contact information. With your contact record with us, you will be provided latest updates and informations regarding various technologies. The format to write us e-mail for resources is as follows:

Subject of E-mail as *Application for resources of .... Black Book.*
You can give your information below to get updates on the book.
Name:
Course pursuing/Profession:
Contact Address:
E-mail ID:

For Any query or suggestion

If you have any query or suggestion please let us know by mailing us on *cadcamcaeworks@gmail.com* or *info@cadcamcaeworks.com*. Your valuable constructive suggestions will be incorporated in our books and your name will be addressed in special thanks area of our books.

About Author

The author of this book, Gaurav Verma, has authored many books on CAD/CAM/CAE topics. He has authored books on **Autodesk Fusion 360, Autodesk Inventor, SolidWorks, SolidWorks Simulation, SolidWorks Flow Simulation, SolidWorks Electrical, AutoCAD Electrical, Creo Parametric 4.0, MasterCAM 2017 for SolidWorks** and many other CAD, CAM, and CAE topics. If you have any query/doubt in any CAD/CAM/CAE package, then you can directly contact the author by writing at cadcamcaeworks@gmail.com.

Chapter 1

Starting with ETABS

The major topics covered in this chapter are:

- *Introduction*
- *Downloading and Installing Educational ETABS*
- *Starting ETABS*
- *Creating a New Model*
- *Steel Design and Concrete Design Codes*
- *User Interface of ETABS*
- *Opening a document*
- *Closing documents*
- *Basic Settings for ETABS*

INTRODUCTION

ETABS is a 3D Modeling and analysis software used to prepare and analyze the building models for real load conditions as per the building design codes. ETABS is capable of performing simple as well as largest & complex non-linear problems of multi-storey. Following are some of the tasks that can be performed easily with the help of ETABS:

- Multi-story commercial, government and health care facilities
- Parking garages with circular and linear ramps
- Buildings with curved beams, walls and floor edges
- Buildings with steel, concrete, composite or joist floor framing
- Projects with multiple towers
- Complex shear walls and cores with arbitrary openings
- Performance based design utilizing nonlinear dynamic analyses
- Buildings based on multiple rectangular and/or cylindrical grid systems
- Flat and waffle slab concrete buildings
- Buildings subjected to any number of vertical and lateral load cases and combinations, including automated wind and seismic loads
- Multiple response spectrum load cases, with built-in input curves
- Automated transfer of vertical loads on floors to beams and walls
- Capacity check of beam-to-column and beam-to-beam steel connections
- P-Delta analysis with static or dynamic analysis
- Explicit panel-zone deformations
- Punching shear checks for concrete slabs
- Construction sequence loading analysis
- Multiple linear and nonlinear time history load cases in any direction
- Foundation/support settlement
- Large displacement analyses
- Nonlinear static pushover
- Buildings with base isolators and dampers
- Design optimization for steel and concrete frames
- Design of concrete slabs using mild reinforcement and post-tensioning
- Design capacity check of steel column base plates
- Floor modeling with rigid or semi-rigid diaphragms
- Automated vertical live load reductions

In this book, we are going to discuss most of these tasks and many others. Now, we will start with downloading and installing ETABS 2016 education version and then we will discuss the user interface of the software.

DOWNLOADING AND INSTALLING ETABS 2016 EVALUATION

The procedure to download and install ETABS 2016 (Evaluation) is given next.

- Open the link https://www.csiamerica.com/support/downloads in your web browser. The CSI Software Download page will be displayed; refer to Figure-1.

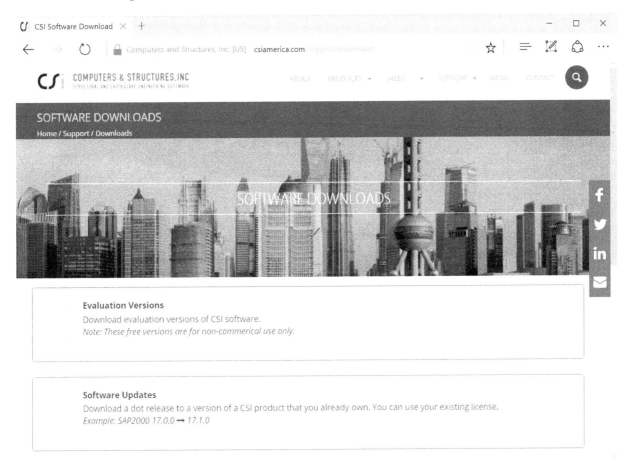

Figure-1. CSI Downloads page

- Click on the **Evaluation Versions** link in the page. Click on the **Request Evaluation** button for **ETABS** in the page; refer to Figure-2. The Software Evaluation Request page will be displayed.

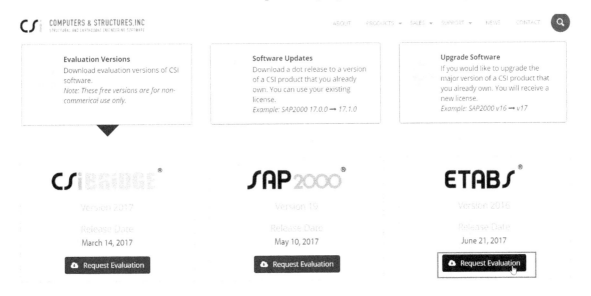

Figure-2. Request Evaluation button for ETABS

- Enter your contact information in the form and click on the **SUBMIT** button. An e-mail will be sent to your provided e-mail id.
- Click on the **Click to Download** link in the e-mail. An executive file will start downloading. Once the file has downloaded, click on it to run.
- Follow the instructions and the software will get installed. Note that there are some restrictions on the uses of evaluation version of the software. If you want to work on live projects then you should buy the professional version of ETABS.

STARTING ETABS

- Click on the **Start** button in Windows 10 and type **ETABS**. List of the programs with the typed characters will be displayed; refer to Figure-3.

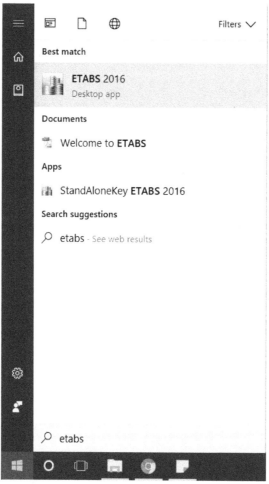

Figure-3. ETABS in Start menu

- Click on the **ETABS 2016** option from the menu. The user interface of ETABS will be displayed; refer to Figure-4.

Figure-4. User interface of ETABS

STARTING A NEW MODEL

- Click on the **New** button from the Toolbar or **New Model** option from the **File** menu or press **CTRL+N** from the keyboard. The **Model Initialization** dialog box will be displayed; refer to Figure-5.

Figure-5. Model Initialization dialog box

- Select the **Use Saved User Default Settings** option if you want to use the default set options saved by you like units, preferences etc. Note that these options are set by using **Save User Default Settings** option in the **Options** menu. We will discuss more about these options later in the book.
- Select the **Use Built-in Settings with** radio button if you want to manually specify the settings related to units, steel database, design codes etc. The option below the radio button will become

active on selecting this radio button. Select the desired parameters in the drop-downs. Details of different drop-downs are given next.

Display Units : The options in this drop-down are used to set units required for various parameters of the model like length, area, angle, temperature, and so on.

Steel Section Database : The options in this drop-down are used to select the desired steel section for use in structural drawing. There are various steel section database codes depending on the country you are in. Like AISC 14 or AISC 14M. More information about these codes can be accessed from related database.

Steel Design Code : The options in this drop-down are used to select the design code for steel as per the requirement.

Concrete Design Code : The options in this drop-down are used to set the desired concrete design code. This code is used to define concrete frame design, shear wall design, and so on.

- Click on the **OK** button from the dialog box to start a new model. The **New Model Quick Templates** dialog box will be displayed; refer to Figure-6.

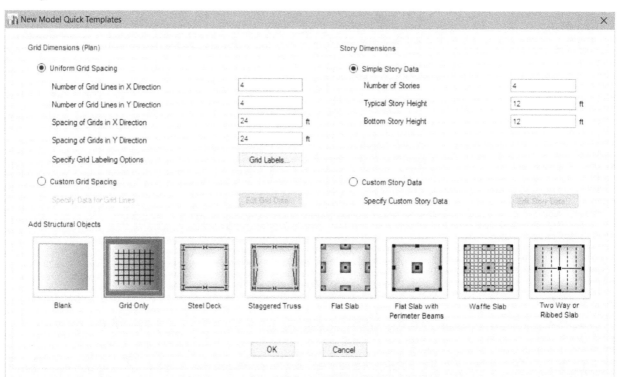

Figure-6. New Model Quick Templates dialog box

Grid Dimensions

- Select the **Uniform Grid Spacing** radio button if you want to create grids with uniform spacing vertically and horizontally, and specify the desired parameters in the edit boxes below it. To set the labels of grid, click on the **Grid Labels** button. The **Grid Labeling Options** dialog box will be displayed; refer to Figure-7. Specify the desired options in the dialog box and click on the **OK** button.

Figure-7. Grid_Labeling_Options_dialog_box.png

- If you want to create custom grid spacing then click on the **Custom Grid Spacing** radio button. The **Edit Grid Data** button will become active. Click on this button. The **Grid System Data** dialog box will be displayed; refer to Figure-8. Set the desired values for grid in the tables. To edit the values, double-click on them in the table. Click on the **OK** button to apply changes. The **New Model Quick Templates** dialog box will be displayed again.

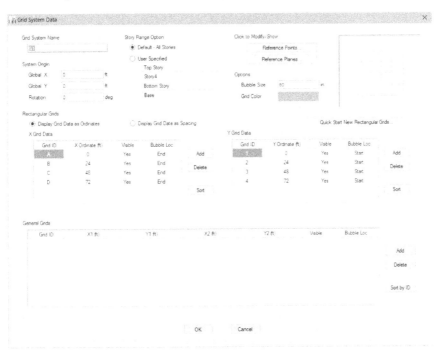

Figure-8. Grid System Data dialog box

Story Dimensions

- In the **Story Dimensions** area of the dialog box, select the **Simple Story Data** radio button to specify the story parameters like number of stories, typical story height, and bottom story height. For example, if you want a 4 story building with bottom height 15 ft and all other story heights 12 ft then specify **4** in **Number of Stories,**

12 in **Typical Story Height,** and **15** in **Bottom Story Height** edit boxes.

- Select the **Custom Story Data** radio button if you want to set the parameter of each story manually. On selecting this radio button, the **Edit Story Data** button will become active. Click on this button. The **Story Data** dialog box will be displayed; refer to Figure-9.

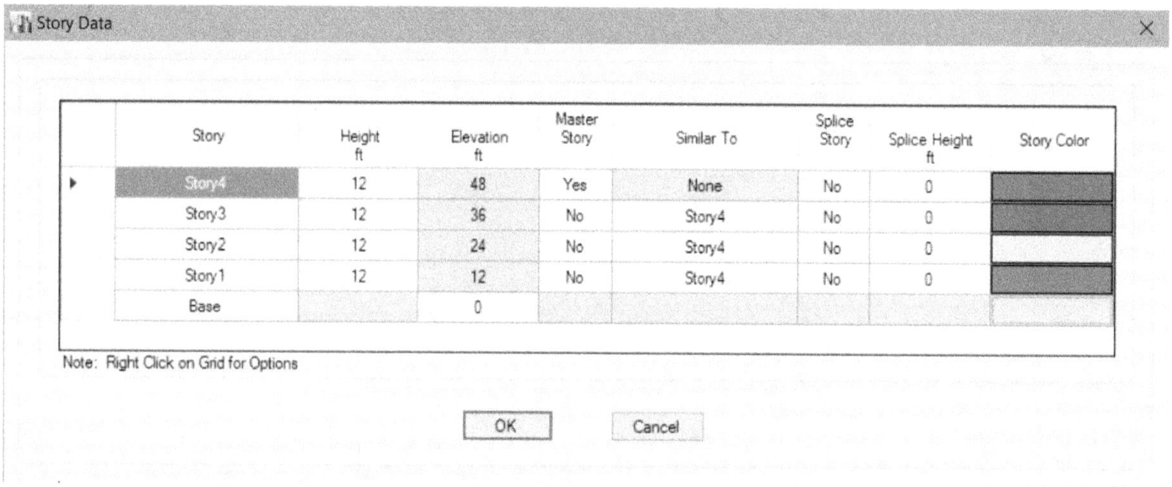

Story	Height ft	Elevation ft	Master Story	Similar To	Splice Story	Splice Height ft	Story Color
Story4	12	48	Yes	None	No	0	
Story3	12	36	No	Story4	No	0	
Story2	12	24	No	Story4	No	0	
Story1	12	12	No	Story4	No	0	
Base		0					

Note: Right Click on Grid for Options

OK   Cancel

Figure-9. Story Data dialog box

- Double-click on the entry in table to change it. Click on the **OK** button to apply the changes. The **New Model Quick Templates** dialog box will be displayed again.

Add Structural Objects

- The buttons in the **Add Structural Objects** area are used to specify the type of structural objects required in template. The description of each button is given next.

Blank Button

Click on the **Blank** button if you want to start a blank model.

Grids Only Button

Click on the **Grids Only** button if you want to display grid lines along with labels in the model.

Steel Deck Button

Click on the **Steel Deck** button if you want to create floors of steel beams & columns. On clicking this button, the **Structural Geometry and Properties for Steel Deck** dialog box will be displayed; refer to Figure-10. In the edit boxes of **Overhangs** area, specify the length up to which the steel desk should overhang from the building columns.

Secondary Beams area

In the **Secondary Beams** area of the dialog box, select the **Secondary Beams** check box to add secondary beams in the model. On selecting this check box, the options below it will become active. Select the desired direction along which the secondary beams are to be created from the **Direction** drop-down. Select the **Max spacing** radio button and specify the maximum spacing to be given between two secondary beams. Select the **Number** radio button and specify the number of secondary beams to be created along the selected direction.

Moment Frame Type area

Select the desired radio button from the **Moment Frame Type** area of the dialog box to create moment resisting frame. If you select the **None** radio button then moment is released at beam-column joints of the model and all the connections are pinned. If you select the **Perimeter** radio button then beam-column connections at the perimeter are moment resisting and all the other connections are pinned. If you select the **Intersecting** radio button then all the column-beam connections are moment resisting. Select the **Special Moment Beams** check box if you want to assign special form to the beam at connections for analysis purpose. After selecting this check box, click on the **Specify Type** button. The **Frame Assignment - Moment Frame Beam Connection Type** dialog box will be displayed; refer to Figure-11.

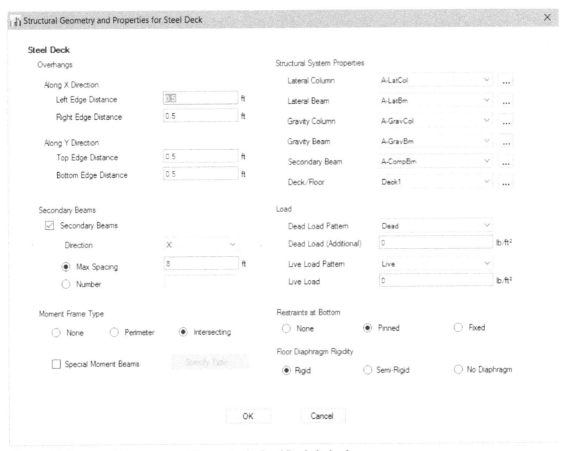

Figure-10. Structural Geometry and Properties for Steel Deck dialog box

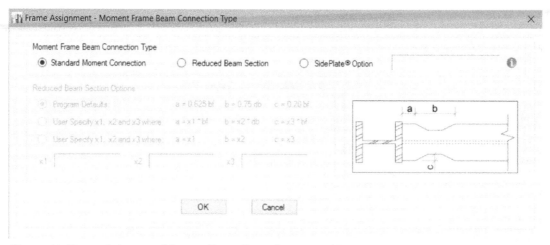

Figure-11. Frame Assignment-Moment Frame Beam Connection Type dialog box

Select the desired connection type from the dialog box and specify the parameters as required. Click on the **OK** button to apply changes.

Structural System Properties area

- Select the desired option from the **Lateral Column** drop-down. The option selected in this drop-down will define frame section property of columns for beam-to-column connections that are fully moment resisting.
- Select the desired option from the **Lateral Beam** drop-down. The option selected in this drop-down will define frame section property of beams for beam-to-column connections that are fully moment resisting.
- Select the desired option from the **Gravity Column** drop-down. The option selected in this drop-down will define frame section property of columns for beam-to-column connections that are pinned.
- Select the desired option from the **Gravity Beam** drop-down. The option selected in this drop-down will define frame section property of beams for beam-to-column connections that are pinned.
- Select the desired option from the **Secondary Beam** drop-down. The option selected in this drop-down will define the frame section property of secondary frame.
- Select the desired option from the **Deck/Floor** drop-down. The option selected in this drop-down will define the frame section property for deck/floor.

Load area

The options in the **Load** area are used to define live/dead loads of the steel desk. These options are discussed next.

- Select the desired option from the **Dead Load Pattern** drop-down. The **Dead** option is by default selected option in this drop-down. The dead load is used to represent the load caused by static structures like beams, columns, floors etc. Specify the desired load value in the **Dead Load (Additional)** edit box.
- Select the desired option from the **Live Load Pattern** drop-down. Live load is used to represent load caused by human movement, furniture, and other moving items. Specify the desired value of load in the **Live Load** edit box below the drop-down.

Restraints at Bottom

The options in this area are used to specify the type of restraints at the bottom of columns of the deck. There are three radio buttons in this area; **None**, **Pinned**, and **Fixed**. Select the **None** radio button if you want no restraints to be applied at the bottom of the columns. Select the **Pinned** radio button if you want to specify pinned restraints at the bottom of the columns. Select the Fixed radio button if you want the columns to be fixed.

Floor Diaphragm Rigidity

The options in this area are used to specify the rigidity level of floor diaphragm. There are three radio buttons in this area. Select the **Rigid** radio button if you want to assume a fully rigid floor diaphragm for analysis purpose. Similarly, you can select **Semi-rigid** radio button for in-plane stiffness of floor. Select the **No Diaphragm** radio button to remove diaphragm rigidity of the floor.

After defining the parameters, click on the **OK** button from the dialog box. The **New Model Quick Templates** dialog box will be displayed again.

Similarly, you can use the other buttons in the **Add Structural Objects** area of the dialog box.

After setting the template parameters, click on the **OK** button from the **New Model Quick Templates** dialog box. A new document will open in the ETABS 2016; refer to Figure-12.

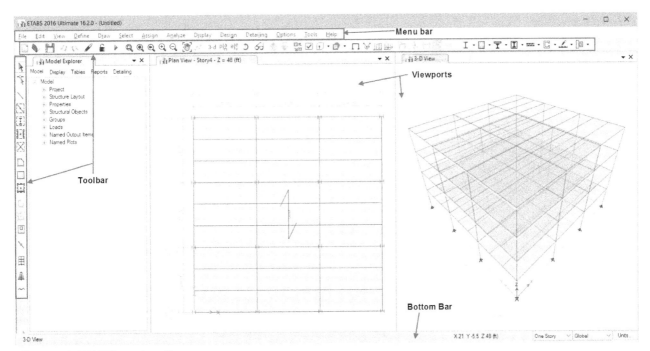

Figure-12. ETABS user interface

USER INTERFACE OF ETABS

As shown in the previous figure, following are the components of user interface of the ETABS:

MENU BAR

Menu bar consists of all the tools required to perform an action in the software. Menu bar contains different menus which have various tools of similar action; refer to Figure-13. Various tools and options in this menu will be discussed later in the book.

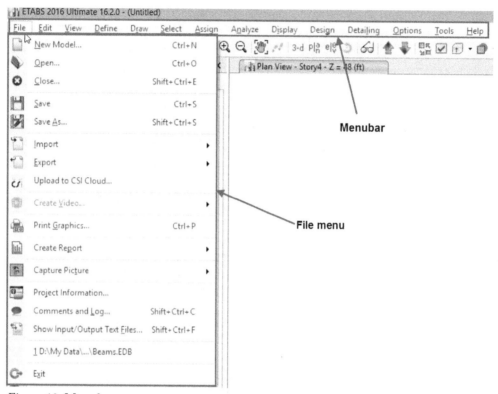

Figure-13. Menu bar

TOOLBAR

Toolbar consists of commonly used tools. Toolbar are available in left and right of the viewports as well as below Menu bar. Tools in the toolbar will be discussed later in the book.

STATUS BAR

Status Bar is used to display information related to floor level.

Working Plane drop-down

From the right-side of the bottom bar, you can use the **Working plane** drop-down to select level at which you want to work.

Story drop-down

Select the desired option from the **Story** drop-down to define which story of the building will be affected by your applied changes. Select the

One Story option from the Story drop-down if you want the changes to be applied only on the selected story. Select the All Stories option from the drop-down if you want to select the elements on all stories of the building. Select the Similar Stories option from the drop-down if you want to select elements from the similar stories.

Grid System drop-down
Select the desired grid system from the Grid System drop-down in the Status bar. You can edit the grid systems by using the Edit Stories and Grid Systems tool from the Edit menu.

Units
The Units button in the Status bar is used to change the unit system of current project. On clicking the Units button, a flyout will be displayed where you can select the desired unit system; refer to Figure-14.

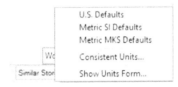

Figure-14. Units flyout

MODEL EXPLORER
Model Explorer is displayed at the left of the viewports. The options in the Model Explorer are used to check and modify the parameters related to model, display, tables, reports, and detailing (drawings). Various tabs of the Model Explorer are discussed next.

Model Tab
The options of the Model tab are used to check and edit various parameters of the model like project preferences, structure layout, loads, and so on; refer to Figure-15.

Figure-15. Model tab in Model Explorer

- Click on the **+** sign before a category in the **Model** tab of the **Model Explorer** to expand the list of parameters under the category.
- To edit the parameter, right-click on it from the **Model** tab and click on the **Edit** button from the shortcut menu. The respective dialog box will be displayed to edit parameter.

Detail about various categories displayed in the **Model** tab is given next.

Project Category

The project category in the **Model** tab is used to edit project information and design preferences. Expand the **Project** category and right-click on the **Project Information** option. A shortcut menu will be displayed. Click on the **Edit** option from the shortcut menu. The **Project Information** dialog box will be displayed; refer to Figure-16.

Figure-16. Project Information dialog box

Click in the desire field to edit the option like click in the **Client Name** edit box and specify the name of your client. Note that information provided here will also be displayed in the final drawing. After specifying the desired parameters, click on the **OK** button from the dialog box.

You can also display the information of project in table form. To do so, right-click on **Project Information** option in the **Project** category of the **Model** tab in the **Model Explorer** and select the **Show Table** option. The **Project Information** box will be displayed at the bottom in the application window; refer to Figure-17.

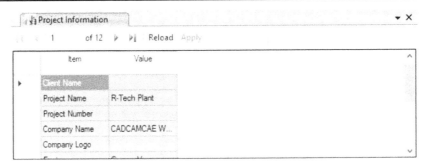

Figure-17. Project Information box

Here, you can check the project information provided by you. Click on the **Close** button (x) at the right corner of the box to close it.

Design Preferences

The options under this node are used to define parameters related to designing frames, beams, and shear walls. To edit any design parameter, expand the **Design Preferences** node and right-click on the design parameter which you want to edit. Like, right-click on the **Steel Frame Design** option and select the **Edit** option from the shortcut menu displayed. The related dialog box will be displayed; refer to Figure-18. To edit the parameters in the table, click in the field and specify the desired value. Click on the **OK** button from the dialog box to apply the changes.

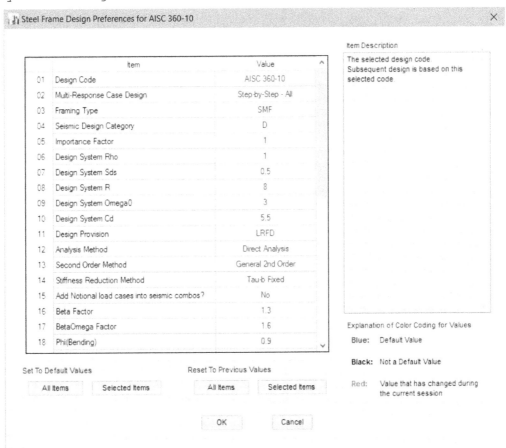

Figure-18. Steel Frame Design Preferences for AISC 360-10 dialog box

Note that the parameters in the above table are directly related to structural designing of building and analysis results so change these values very carefully. Details of each parameter can be found in your civil engineering books.

Structure Layout Category

Structure Layout consists of stories and grid system defined by you. The procedure to edit various parameters of structural layout is given next.

- Expand the nodes of **Structure Layout** to check various parameters defined for the structure layout. To edit the layout, right-click on it. The **Edit Story and Grid Systems** button will be displayed.
- Click on the **Edit Story and Grid Systems** button, the **Edit Story and Grid System Data** dialog box will be displayed; refer to Figure-19.

Figure-19. Edit Story and Grid System Data dialog box

- To edit story data, click on the **Modify/Show Story Data** button. The **Story Data** dialog box will be displayed; refer to Figure-20.
- Edit the story data as required and click on the **OK** button. The **Edit Story and Grid System Data** dialog box will be displayed again. You will learn more about these options later in the book.
- To modify existing grid, select the grid from the **Grid Systems** list box and click on the **Modify/Show Grid System** button. The **Grid System Data** dialog box will be displayed; refer to Figure-21.

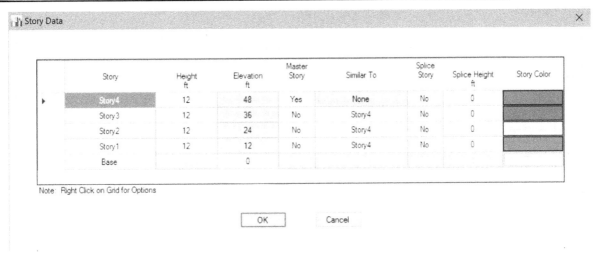

Figure-20. Story Data dialog box

Figure-21. Grid System Data dialog box

- Set the parameters as required and click on the **OK** button. Note that you will learn more about this dialog box later in this book.
- Similarly, you can use the other options of this dialog box. Click on the **OK** button from the **Edit Story and Grid System Data** dialog box.

Similarly, you can check and edit the other parameters of **Model** tab in the **Model Explorer**.

Display Tab

- Click on the **Display** tab in the **Model Explorer** to check various views of the model in viewports; refer to Figure-22. The options in this tab are used to edit the parameters related to views.

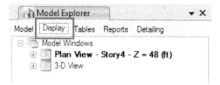

Figure-22. Display tab of Model Explorer

- By default there are two views created in Model Windows (Viewports) viz. Plan View and 3-D View. Expand the view that you want to edit by clicking on the **+** button. Four categories of the view are displayed viz. View, Limits, Options, and Display; refer to Figure-23.

- Expand the **View** node and select the desired radio button to change view to **3D**, **Plan**, or **Elevation**.

Figure-23. Options for Views in Display tab

- Expand the **Limits** node and click on the **Set Limits** option. The **Set Building View Limits** dialog box will be displayed; refer to Figure-24. Set the desired minimum and maximum limits of the view and click on the **OK** button to apply changes.

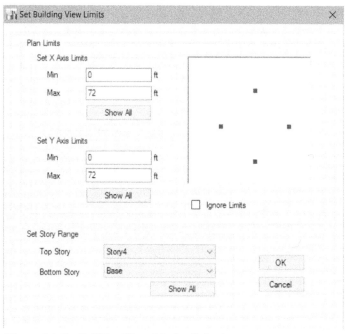

Figure-24. Set Building View Limits dialog box

Expand **Options** node in the **Display** tab of **Model Explorer**. The options to hide/show different elements of model are displayed. Expand the desired category and select/de-select the check boxes to show/hide the elements of the model.

Expand the **Display** node and select the desired radio button to display deformed and undeformed shape.

Tables Tab

The options in the **Tables** tab of the **Model Explorer** are used to show and export tables; refer to Figure-25.

- Expand the desired category of **Tables** node and right-click on the table option that you want to be displayed; refer to Figure-26.

Figure-25. Tables tab of Model Explorer

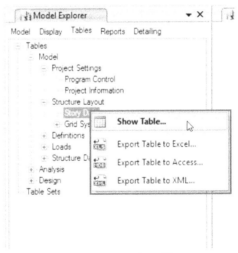

Figure-26. Shortcut menu for table

- Click on the **Show Table** button if you want to check the table of parameters for selected element in the **Model Explorer**. A box will be displayed containing the table; refer to Figure-27.

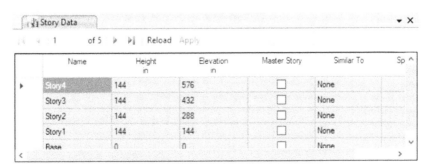

Figure-27. Story Data table box

Similarly, you can check the other tables of the model.

If you want to export the table to Excel, Access or XML format then click on the respective button from the shortcut menu like select the **Export to Excel** option if you want to export the table in excel format.

Reports Tab

The **Reports** tab contains all the data of project compiled in report forms; refer to Figure-28. To generate the project report, right-click on **Project Report** category in **Reports** node of the **Reports** tab of **Model Explorer** and click on **Show/Refresh Report** option; refer to Figure-29. A report will be generated and displayed in the viewport; refer to Figure-30.

Figure-28. Reports tab of Model Explorer

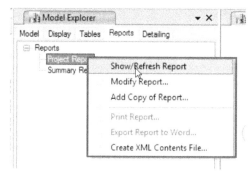

Figure-29. Project Report shortcut menu

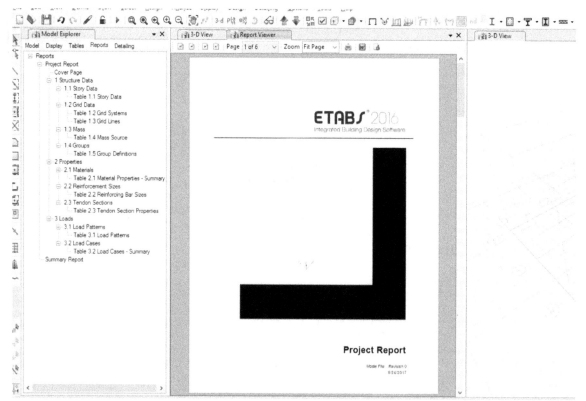

Figure-30. Project Report

Detailing Tab

The options in the **Detailing** tab are used to describe the level of details for the model. You can find the same options in the **Detailing** menu. We will be using the parameters of the detailing tab while working the succeeding chapters.

FILE MENU

The tools in the **File** menu are used to manage file level operations like creating a new file, opening an existing ETABS file, importing or exporting files, and so on. Various functions of the **File** menu are discussed next.

Creating a New Model

The **New Model** tool in the **File** menu is used to create a new ETABS model. The procedure to start a new model has been discussed earlier.

Opening a Model File

The **Open** tool in the **File** menu is used to open existing models. You can also use the shortcut key **CTRL+O** to open a model file. The procedure is given next.

- Click on the **Open** tool from the **File** menu. The **Open Model File** dialog box will be displayed; refer to Figure-31.

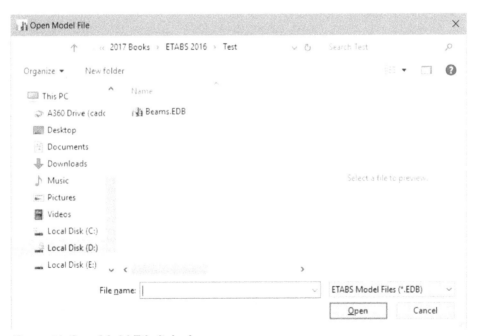

Figure-31. Open Model File dialog box

- Browse to the folder in which the file is available that you want to open and double-click to open it. Note that the extension for Etabs model file is ".EDB".

Closing a Model File

The **Close** tool in the **File** menu is used to close the model on which you are currently working. You can also use the shortcut key **SHIFT+CTRL+E** to close the model. If your file is not saved after making changes then you will be asked to save the changes via a dialog box.

Saving a File

You can save the file using **Save** and **Save As** tools in the **File** menu to save the model file. Using the **Save** tool, you can update the changes in the current file if you have performed any changes. Using the **Save As** tool, you can save the current file with different name or format. If you are using the **Save** tool for the first time then it will function similar to **Save As** tool. The procedure to save a file is given next.

* Click on the **Save** tool after making the desired changes. The **Save Model File As** dialog box will be displayed (refer to Figure-32) if you are saving the file for the first time.

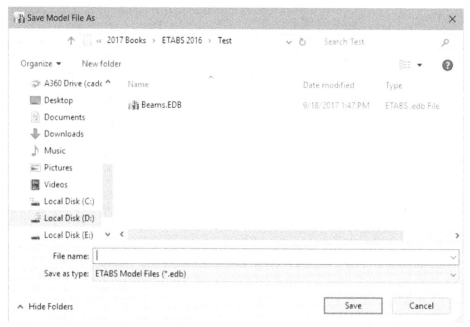

Figure-32. Save Model File As dialog box

* Specify the desired name and click on the **OK** button to save the file.
* If you have already saved the file and then made some changes, clicking on the **Save** button will save the file without showing any dialog box.
* Similarly, you can use the **Save As** tool in the **File** menu to save the file in different format or using the different name.

Note that when you save the file then *.ebk file is also created as backup for the current file.

Importing Files/Data

The options to import files or data in the model are available in the **Import** cascading menu of the **File** menu; refer to Figure-33.

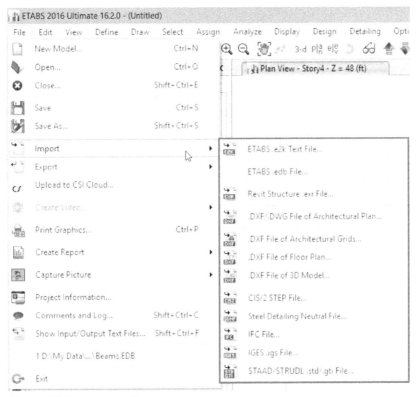

Figure-33. Import cascading menu

The procedures to import different files are given next.

Importing ETABS .e2k Text File

There is one more file type associated with '.e2k' file format viz. '*.$et'. The .$et file is created every time you save a model in ETABS. This file contains the text commands for the model; refer to Figure-34. If you have worked with earlier versions of AutoCAD which worked on **Command Manager** only then you may find editing these files very interesting.

```
$ PROGRAM INFORMATION
  PROGRAM  "ETABS 2016"  VERSION "16.2.0"

$ CONTROLS
  UNITS  "LB"  "IN"  "F"
  TITLE2  "Beams"
  PREFERENCE  MERGETOL 0.1
  RLLF  METHOD "ASCE7-10"  USEDEFAULTMIN "YES"

$ TOWERS
  TOWER "T1"

$ STORIES - IN SEQUENCE FROM TOP - FOR EACH TOWER
  STORY "Story4"  HEIGHT 144 MASTERSTORY "Yes"  TOWER "T1"
  STORY "Story3"  HEIGHT 144 SIMILARTO "Story4"
  STORY "Story2"  HEIGHT 144 SIMILARTO "Story4"
  STORY "Story1"  HEIGHT 180 SIMILARTO "Story4"
  STORY "Base"  ELEV 0

$ GRIDS
  GRIDSYSTEM "G1"  TOWER "T1"  TYPE "CARTESIAN"  BUBBLESIZE 60
  GRID "G1"  LABEL "A"  DIR "X"  COORD 0 VISIBLE "Yes"  BUBBLELOC
"End"
  GRID "G1"  LABEL "B"  DIR "X"  COORD 288 VISIBLE "Yes"
BUBBLELOC "End"
  GRID "G1"  LABEL "C"  DIR "X"  COORD 576 VISIBLE "Yes"
BUBBLELOC "End"
  GRID "G1"  LABEL "D"  DIR "X"  COORD 864 VISIBLE "Yes"
BUBBLELOC "End"
  GRID "G1"  LABEL "1"  DIR "Y"  COORD 0 VISIBLE "Yes"  BUBBLELOC
"Start"
  GRID "G1"  LABEL "2"  DIR "Y"  COORD 288 VISIBLE "Yes"
```

Figure-34. Text commands in $et file

The procedure to import *.e2k or *.$et file is given next.

- Click on the **ETABS .e2k Text File** option from the **Import** cascading menu in the **File** menu. The **Open ETABS Text File** dialog box will be displayed; refer to Figure-35.

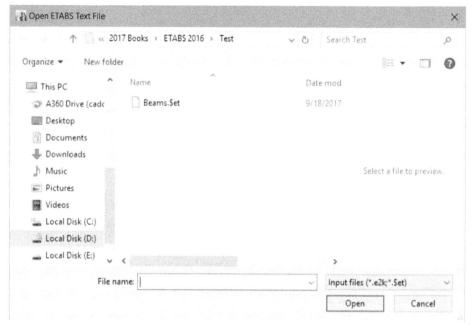

Figure-35. Open ETABS Text File dialog box

- Double-click on the ETABS text file that you want to import. The model will be generated based on your specified command text file.

Importing ETABS .edb File

The .edb is binary file saved by ETABS to store all the data related to your model. The procedure to import this file is similar to previous topic discussed.

Importing Revit Structure Exchange File

Revit Structure is one of the most widely used structural designing software by Autodesk. You can import the exchange file generated in Revit Structure by following steps.

- Click on the **Revit Structure .exr File** option from the **Import** cascading menu in the **File** menu. The **Import ETABS/Revit Structure Exchange File** dialog box will be displayed; refer to Figure-36.

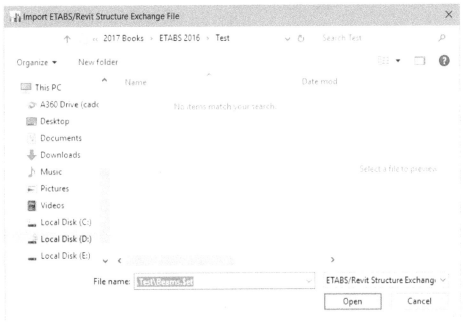

Figure-36. Import Etabs/Revit Structure Exchange File dialog box

- Double-click to open the desired *.EXR file.

Importing DXF/DWG Files

There are four options in the **Import** cascading menu to import '.DXF' and '.DWG' files in ETABS:

- .DXF/.DWG File of Architectural Plan
- .DXF File of Architectural Grid
- .DXF File of Floor Plan
- .DXF File of 3D Model

Select the '**.DXF/.DWG File of Architectural Plan**' option from the **Import** cascading menu if you want to import a dxf or dwg file which contains architectural plan of a building. Select the '**.DXF File of Architectural Grid**' option if you want to import only architectural grid of the building. Select the '**.DXF File of Floor Plan**' option if you want to import the floor plans from other architectural designing software. Select the '**.DXF File of 3D Model**' option to import any 3D model for other CAD software. The procedure to import .dxf file of architectural plan is given next. You can use the same procedure with other options.

- Click on the **.DXF/.DWG File of Architectural Plan** option from the **Import** cascading menu in the **File** menu. The **Import DXF/DWG File** dialog box will be displayed; refer to Figure-37.

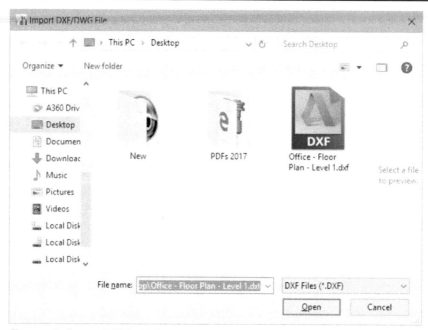

Figure-37. Import DXF DWG File dialog box

- Select the file that you want to import and click on the **Open** button. The **Architectural Plan Import** dialog box will be displayed; refer to Figure-38.

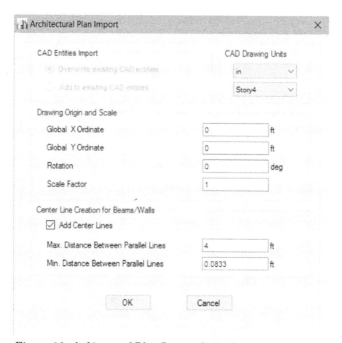

Figure-38. Architectural Plan Import dialog box

- Set the desired parameters in the dialog box like units, level in ETABS, drawing origin, and so on.
- Click on the **OK** button. The drawing will be placed in the current ETABS model.

Similarly, you can use the other options in the **Import** cascading menu to import different type of files.

Exporting Files

Like importing files, there are various options to export the files. These options are available in the **Export** cascading menu of the **File** menu; refer to Figure-39.

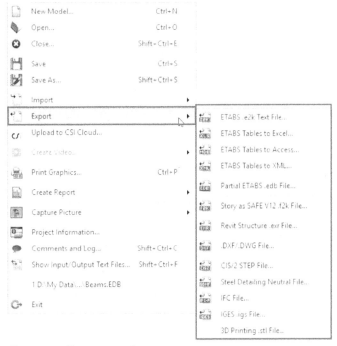

Figure-39. Export cascading menu

The procedure to export .exr file for Revit Structure is discussed next. You can apply the same procedure to other options.

- Click on the **Revit Structure .exr File** option from the **Export** cascading menu. The **Export ETABS/Revit Structure Exchange File** dialog box will be displayed; refer to Figure-40.

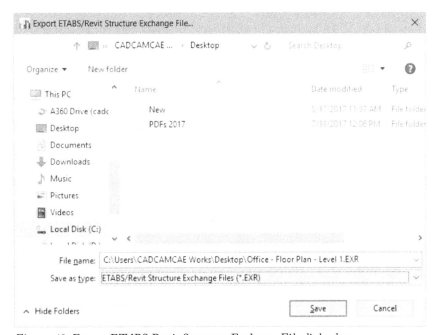

Figure-40. Export ETABS Revit Structure Exchange File dialog box

- Specify the name of file by which you want to save it in the **File name** edit box and click on the **Save** button. The file will be exported in the specified format.

Upload to CSI Cloud

The **Upload to CSI Cloud** option in the **File** menu is used to upload the files on cloud.

Creating Video of Analysis

Once you have performed desired structural analyses, you can record a video of results to present in front of your customers. The tools to create videos are available in the **Create Video** cascading menu in the **File** menu; refer to Figure-41. Note that these options are active only after you have run the analysis on model. There are two options in the cascading menu: **Multi-step Animation** and **Cycle Animation**. Select the **Multi-step Animation** option if you have run a time history analysis and want to record the effect of loads at different time steps. Select the **Cycle Animation** option if you want to record the effect of loads on the model in cycle.

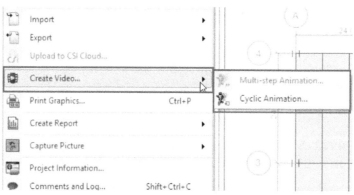

Figure-41. Create Video cascading menu

The procedure to create video of cyclic animation is given next. You can use the same procedure to create **Multi-step Animation**.

- After running the analysis, click on the **Cyclic Animation** option from the **Create Video** cascading menu. The **Video File** dialog box will be displayed; refer to Figure-42.
- Specify the desired name of the file in the **File name** edit box and click on the **Save** button. The **Animation Video File Creation** dialog box will be displayed; refer to Figure-43.

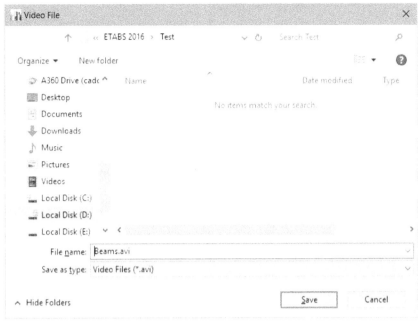

Figure-42. Video File dialog box

Figure-43. Animation Video File Creation dialog box

- Specify the desired parameters in the dialog box and click on the **OK** button. The animation will run once and the video file will be saved at specified location.

Printing Graphics

The **Print Graphics** tool in the **File** menu is used to print the object in the selected viewport. The procedure to print graphics is given next.

- Click on the **Print Graphics** tool from the **File** menu or press **CTRL+P** from keyboard. The **Print Graphics** window will be displayed; refer to Figure-44.

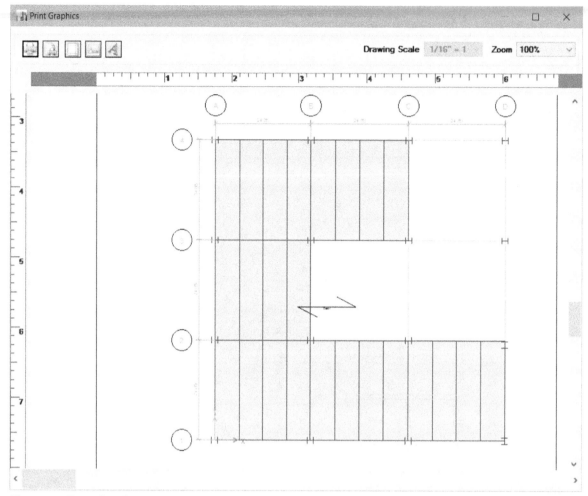

Figure-44. Print Graphics window

- There are various buttons in the top left corner of the window to perform printing and editing operations. Following are the operations that you can perform while in the **Print Graphics** window:

Adding Note to the Prints

- Click on the **Add Note** button from the **Print Graphics** window and click at the desired location to place the text note. You will be asked to specify the desired text and a toolbar will be displayed to specify font parameter.
- Specify the desired parameter and type the text as desired.
- Click on the **Done** button to display text. You can drag this text at desired location.
- If you want to edit this text then double-click on it. The text will become edit-able and the toolbar will be displayed; refer to Figure-45. Modify the text as required and click on the **Done** button.

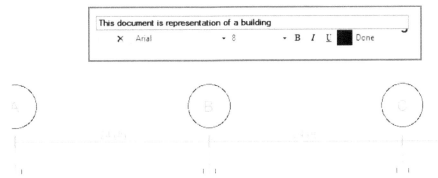

Figure-45. Editing text of note

Adding Image to Print

The **Add Image** button in the **Print Graphics** window is used to insert an image in the print. Follow the steps given next to add image.

- Click on the **Add Image** button from **Print Graphics** window and click at the desired location. The **Open** dialog box will be displayed; refer to Figure-46.

Figure-46. Open dialog box

- Select the desired image file and click on the **Open** button. The image will be placed at specified location.
- Set the desired parameters in the toolbar displayed; refer to Figure-47 and click on the **Done** button.

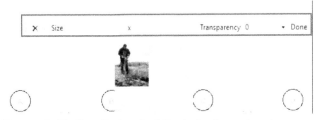

Figure-47. Toolbar displayed while placing image in print

Page Setup for Printing

- Click on the **Page Setup** button from the **Print Graphics** window. The **Page Setup** dialog box will be displayed; refer to Figure-48.

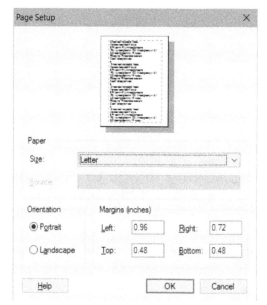

Figure-48. Page Setup dialog box

- Select the page size and specify the other parameters as required.
- Click on the **OK** button after specifying the parameters. The page setup for printing will get changed.

Setting Printer

- Click on the **Printer Setup** button from the **Print Graphics** window. The **Print** dialog box will be displayed.
- Select the printer by which you want to print the current objects.
- Set the number of copies and other related parameters as required.
- Click on the **OK** button from the dialog box to apply changes.

Printing the Model

- After setting all the required parameters in the **Print Graphics** window, click on the **Print** button. The file will be printed by selected printer.

Generating Reports

The tools to generate reports are available in the **Create Report** cascading menu of the **File** menu; refer to Figure-49.

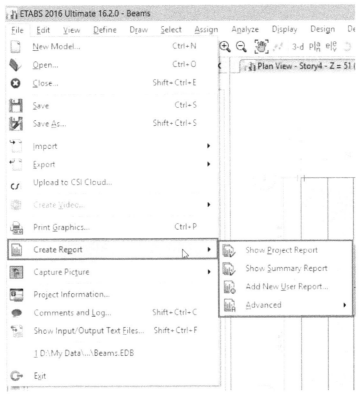

Figure-49. Create Report cascading menu

Various tools in this cascading menu are discussed next.

Show Project Report Tool

- Click on the **Show Project Report** tool from the **Create Report** cascading menu. The **Report Viewer** will be displayed with automatically generated report; refer to Figure-50.
- Click on the **Print** button in the **Report Viewer** if you want to take a printout of the report.
- Click on the **Export to Word Document** 📄 button if you want to export the report in MS Word format.
- Click on the **Regenerate Report** 🔄 button if you want to update the report after making changes in the model.

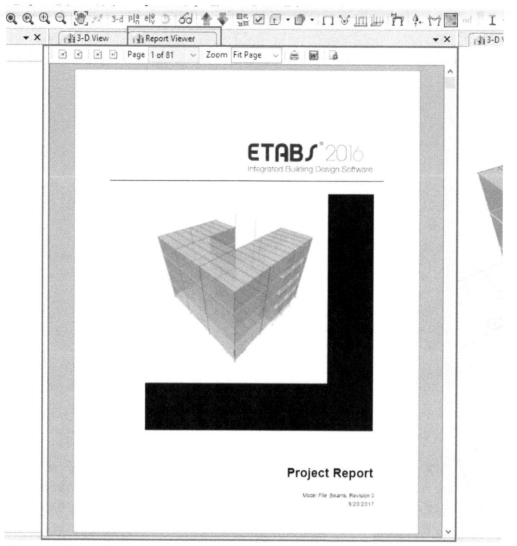

Figure-50. Report Viewer

- To close the report, click on the '**x**' button at the top right corner of the **Report Viewer**.

Similarly, you can use the **Show Summary Report** tool in the **Create Report** cascading menu.

Add New User Report Tool

The **Add New User Report** tool is used to generate the report based on user defined parameters. The procedure to use this tool is given next.

- Click on the **Add New User Report** tool from the **Create Report** cascading menu in the **File** menu. The **User Report** dialog box will be displayed; refer to Figure-51.

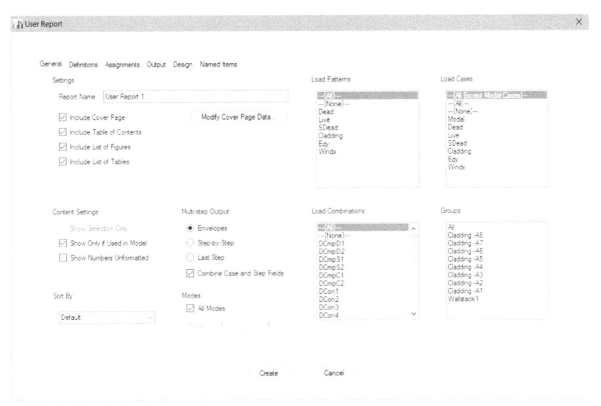

Figure-51. User Report dialog box

- Select the check boxes for all the parameters that you want to be displayed in the report. Note that there are 6 tabs for different categories in the dialog box.
- After setting the parameters, click on the **Create** button. The report will be generated.

Capturing Picture

The tools to take a screen shots are available in the **Capture Picture** cascading menu of the **File** menu; refer to Figure-52.

Select the desired option from the cascading menu to take snapshot. Like, select the **Main Window** option from the cascading menu (or press **Shift**+**Ctrl**+**6**) to take a snapshot of the whole screen.

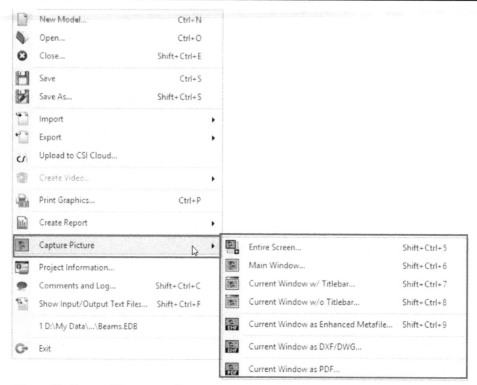

Figure-52. Capture Picture cascading menu

Checking the Project Information

The **Project Information** tool is used to check and update the information related to a project. The procedure to check project information is given next.

- Click on the **Project Information** tool from the **File** cascading menu. The **Project Information** dialog box will be displayed; refer to Figure-53.

Figure-53. Project Information dialog box

- Specify the information as required in the dialog box as discussed earlier.
- Click on the **OK** button to apply changes.

Using the **Comments and Log** tool of the **File** menu, you can check the log of operations generated by software related to file handling. This log is useful when there is any problem in loading the model.

Using the **Show Input/Output Text Files** tool, you can check the input and output files related to your model.

FOR STUDENT NOTES

Chapter 2

Starting a Project in ETABS

Topics Covered

The major topics covered in this chapter are:

- *Workflow in ETABS*
- *Modifying Grid*
- *Drawing Joint Objects*
- *Drawing Grid Lines Manually*
- *Snap Options*
- *Drawing Beams, Columns, and Brace Objects*
- *Drawing Floor and Wall Objects*
- *Drawing Links*

- *Drawing Dimension Lines*
- *Drawing Reference Points*
- *Setting Views in Viewports*
- *Drawing Reference Planes*
- *Drawing Developed Elevation Definition*
- *Drawing Wall Stack*
- *Auto Draw Cladding*

INTRODUCTION

In the previous chapter, you have learned about basic operations required for starting a model project in ETABS. In this chapter, you will learn to draw different types of components of building. The tools to draw different components are available in various menus of the ETABS interface. The workflow to create design in ETABS is given next.

WORKFLOW IN ETABS

Figure-1 below shows the common workflow used for structural designing by ETABS.

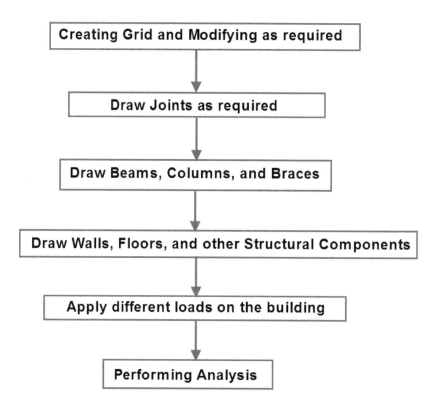

Figure-1. Workflow in ETABS

Now, we will discuss the tools as per the workflow. Start a new document with Grid lines.

MODIFYING GRID

The tools to edit/modify grid data are available in **Edit** menu. Various procedures to edit/modify data are given next.

Setting Grid Options

The options for grid setting are available in the **Grid Options** cascading menu viz. **Glue Joints to Grids** and **Lock Onscreen Grid System Edit**. These options are discussed next.

- Click on the **Glue Joints to Grids** option from the **Grid Options** cascading menu to "glue" joints along a grid line so that if the grid line is moved then the joints move with it.
- Select the **Lock Onscreen Grid System Edit** option from the **Grid Options** cascading menu to enable or disable editing of grid points in the screen.

Editing Stories and Grid System

The **Edit Stories and Grid System** option in the **Edit** menu is used to edit the data related to grids and stories. On clicking this option, the **Edit Story and Grid System Data** dialog box will be displayed; refer to Figure-2.

Figure-2. Edit Story and Grid System Data dialog box

The options of this dialog box have already been discussed in previous chapter.

DRAWING JOINT OBJECTS

The joints are used to locate beams and columns in the structure. The procedure to create joint is given next.

- Click on the **Draw Joint Objects** tool from the **Draw** menu. You will be asked to specify the location of joint.
- Click at the desired location in the structure to define joint location. The joint will be created; refer to Figure-3.

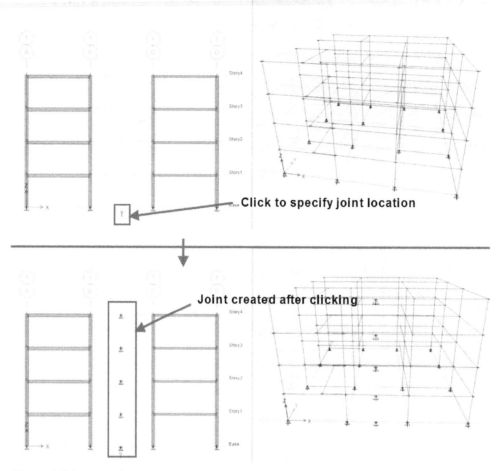

Figure-3. Joint created

Adding Grid Lines at Selected Joints

The **Add Grid Lines at Selected Joints** tool is used to add grid lines at the joints created in structure as discussed in previous topic. The procedure to add grid lines is given next.

• Click on the **Add Grid Lines at Selected Joints** tool from the **Edit** menu. The **Add Grid Lines at Selected Joints** dialog box will be displayed; refer to Figure-4.

Figure-4. Add Grid Lines at Selected Joints dialog box

- Select the grid system in which you want to add the newly created grid from the **Grid System** drop-down.
- Select the check box(es) to define direction of the grid lines, like select the **Parallel to Y** check box to create grid lines parallel to Y direction.
- Click at the joints from which you want to create the grid lines. Click on the **OK** button from the dialog box to create the grid lines; refer to Figure-5.

Figure-5. Grid lines created

DRAWING GRID LINES MANUALLY

In previous topics, we have discussed about creating grid lines automatically based on specified joints. In this section, we will discuss the procedure to create grid lines manually.

- Click on the **Draw Grids** tool from the **Draw** menu. You will be asked to select joints for creating grid lines.
- Click at the desired joints one by one to create grid lines; refer to Figure-6.

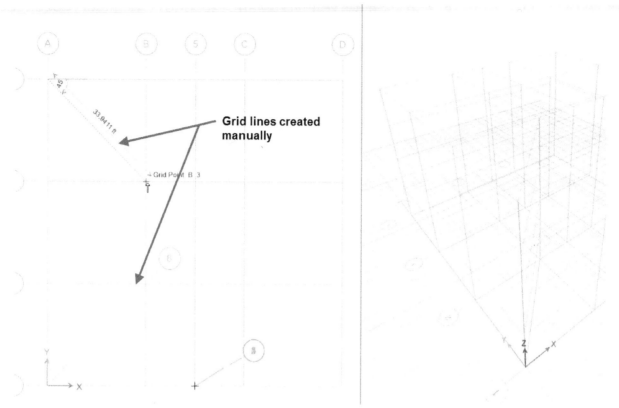

Figure-6. Grid lines created manually

SNAP OPTIONS

Snap options are used to restrict movement of cursor at some special points like mid points, end points, and other points. The procedure to modify which points to be snapped is given next.

Modifying Snap Options

- Click on the **Snap Options** tool from the **Draw** menu. The **Snap Options** dialog box will be displayed; refer to Figure-7.
- Select the check boxes for major points to which you want the cursor to snap.
- Similarly, set the desired parameters as required and click on the **Apply** button.
- Click on the **Close** button to exit the dialog box.

Snap Options   x

Snap to   Settings

☑ Joints   ☐ Intersections   Plan Fine Grid Spacing   12   in

☐ Line Ends and Midpoints   ☐ Fine Grid   Plan Nudge Value   12   in

☑ Grid Intersections   ☐ Extensions   Screen Selection Tolerance 5   pixels

☐ Lines and Frames   ☐ Parallels   Screen Snap To Tolerance   5   pixels

☐ Edges   ☑ Intelligent Snaps   ☐ Drawing Scale   1/16" = 1 ft

☐ Perpendicular Projections ☑ Arch. Layer

        Select All   Deselect All

Snap Increments

☑ (Imperial in Inches) Snap at length increments of
12, 6, 1, 0.25.

☑ (Metric in mm) Snap at length increments of
500; 100; 25; 5.

☑ (Degree) Snap at angle increments of
1;

                          Apply   Close

Figure-7. Snap Options dialog box

Enabling/Disabling Snapping Function
- Click on the **Snap to Grid Intersections & Points** button at the bottom-left of the screen to enable/disable.

Drawing with Snap Only
The **Draw Using Snap Only** tool in the **Draw** menu is used to draw objects in ETABS using the snap points only. To enable or disable this function, click on the **Draw Using Snap Only** option from the **Draw** menu.

DRAWING BEAMS/COLUMNS/BRACE OBJECTS
The tools to draw beams, columns, and brace objects are available in the **Draw Beam/Column/Brace Objects** cascading menu; refer to Figure-8.

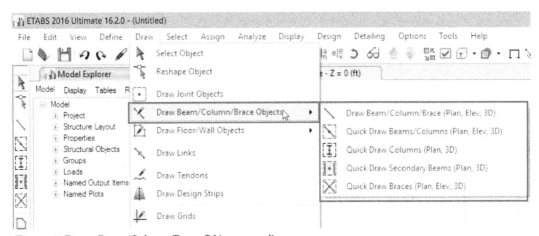

Figure-8. Draw Beam/Column/Brace Objects cascading menu

Various tools in this cascading menu are discussed next.

Drawing Beams/Columns/Braces in Plan, Elev, and 3D

The **Draw Beams/Columns/Braces (Plan, Elev, 3D)** tool is used to draw beams, columns, and braces in plan, elevation, and 3D views. The procedure to use this tool is given next.

- Click on the **Draw Beams/Columns/Braces (Plan, Elev, 3D)** tool from the **Draw Beam/Column/Brace Objects** cascading menu. The **Properties of Object** form will be displayed in the **Model Explorer**; refer to Figure-9 and you will be asked to specify the grid line to create beams, columns, or braces.

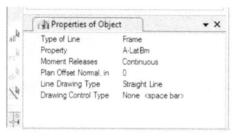

Figure-9. Properties of Object form

- Select the type of line from the **Type of Line** drop-down in the form.
- Click on the down button in the **Property** drop-down of the form and select the desired section type from the list. The option selected in this drop-down will define the material and shape of beam/column/brace.
- Select the desired moment release option from the **Moment Releases** drop-down. There are two options in this drop-down: **Continuous** and **Pinned**. Select the **Continuous** option if the beam is in continuous connection with the connecting beam. Select the **Pinned** option if the beam/column/brace is pin joined with the connecting beam/column.
- Select the offset from current plane as required if you have selected the **Straight Line** option from the **Line Drawing Type** drop-down.
- Select the desired option from the **Line Drawing Type** drop-down. There are six options in the drop-down to draw different shapes: **Arc (3 Points)**, **Arc (Center & 2 Points)**, **Bezier**, **Multilinear**, **Spline**, and **Straight Line**. Based on your selection, you will be asked to specify the desired points to create the shape.
- Specify the points in the desired view to create beam/column/brace. Like if you want selected the **Arc (3 Points)** option then specify the start point, end point, and middle point of arc to create the object; refer to Figure-10.

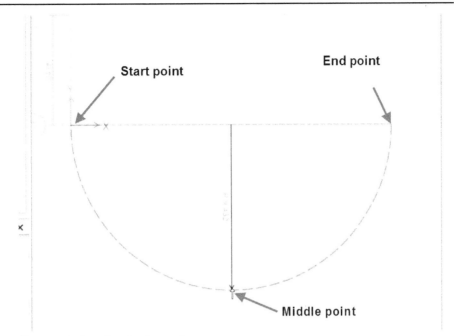

Figure-10. Creating 3 point arc beam

- Press **ESC** to exit the tool.

Note that by default, the object is created only on one story selected in Plan View. If you want the same object to be created on all stories then select the **All Stories** option from the **Story** drop-down at the bottom right corner of the software; refer to Figure-11 while creating the object.

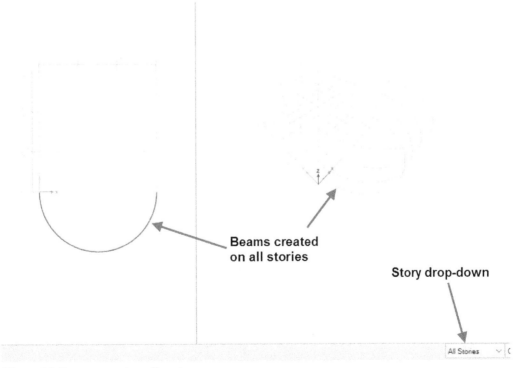

Figure-11. Beams created on all stories

Quick Draw Beams/Columns (Plan, Elev, 3D)

The **Quick Draw Beams/Columns (Plan, Elev, 3D)** tool is used to quickly draw beams/columns using grid lines and points of grids. The procedure to use this tool is given next.

- Click on the **Quick Draw Beams/Columns (Plan, Elev, 3D)** tool from the **Draw Beam/Column/Brace Objects** cascading menu in the **Draw** menu. The **Properties of Object** form will be displayed at the bottom in the **Model Explorer** and you will be asked to select grids.
- Select the grids from the desired viewport. The beams/columns will be created accordingly. The options in the **Properties of Object** form are same as discussed earlier.

Similarly, you can use the **Quick Draw Columns (Plan, 3D)** tool to create columns using Grids or grid points.

Quick Draw Secondary Beams (Plan, 3D)

The **Quick Draw Secondary Beams (Plan, 3D)** tool is used to draw secondary beams which transfer their loads on primary beams. The procedure to use this tool is given next.

- Click on the **Quick Draw Secondary Beams (Plan, 3D)** tool from the **Draw Beam/Column/Brace Objects** cascading menu in the **Draw** menu. You will be asked to specify the location of secondary beams and the **Properties of Object** form will be displayed at the bottom in the **Model Explorer**.
- Select the desired option from the **Spacing** drop-down and enter the desired value in the edit box below it. For example, if you have selected the **No. of Beams** option from the **Spacing** drop-down then specify the number of secondary beams required along the grid.
- Set the desired orientation in the **Approx. Orientation** drop-down.
- Click at the desired location in the grids to place the secondary beams; refer to Figure-12.

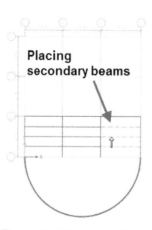

Figure-12. Placing secondary beams

Quick Draw Braces (Plan, Elev, 3D)

The **Quick Draw Braces (Plan, Elev, 3D)** tool is used to draw braces in the structure. Braces are used to increase the stability of structure in case of earth quakes or other dynamic loading cases. The procedure to draw braces in the structure is given next.

- Click on the **Quick Draw Braces (Plan, Elev, 3D)** tool from the **Draw Beam/Column/Brace Objects** cascading menu of **Draw** menu. You will be asked to specify the location of braces.
- Click in the desired grid box to create braces; refer to Figure-13.

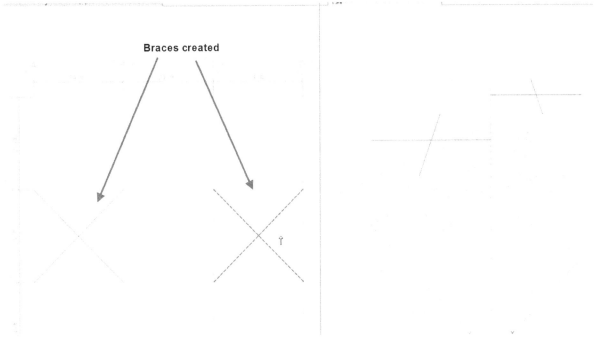

Figure-13. Braces created

DRAWING FLOOR AND WALL OBJECTS

The tools to draw floor and wall objects are available in the **Draw Floor/Wall Objects** cascading menu of the **Draw** menu; refer to Figure-14. Various tools of this cascading menu are discussed next.

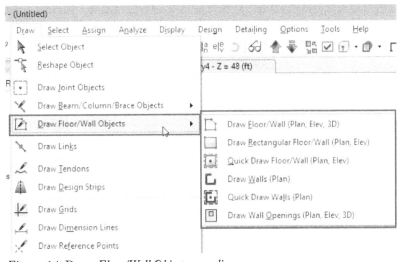

Figure-14. Draw Floor/Wall Objects cascading menu

Draw Floor/Wall (Plan, Elev, 3D) Tool

The **Draw Floor/Wall (Plan, Elev, 3D)** tool is used to draw floors and walls in plan, elevation and 3D view by using the grid points. The procedure to use this tool is given next.

- Click on the **Draw Floor/Wall (Plan, Elev, 3D)** tool from the **Draw Floor/ Wall Objects** cascading menu of the **Draw** menu. You will be asked to select the grid point and the **Properties of Object** form will be displayed.
- Select the desired option from the **Property** drop-down. The options in this drop-down are Slab, Opening, Plank, Deck, and Opening.
- Select the desired options from the **Edge Drawing Type** and **Drawing Control Type** drop-downs as discussed earlier.
- After specifying corner points of the floor/wall, press **ENTER** to create floor/wall; refer to Figure-15.

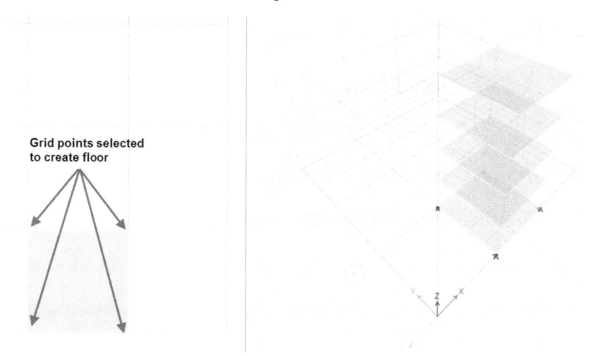

Grid points selected to create floor

Figure-15. Grid points selected to create floor

Draw Rectangular Floor/Wall (Plan, Elev) Tool

The **Draw Rectangular Floor/Wall (Plan, Elev)** tool is used to draw rectangular floor and wall in plan and elevation views. The procedure to use this tool is given next.

- Click on the **Draw Rectangular Floor/Wall (Plan, Elev)** tool from the **Draw Floor/Wall Objects** cascading menu of the **Draw** menu. You will be asked to select the grid point and the **Properties of Object** form will be displayed.
- Click at the desired grid point and drag the cursor while holding LMB to draw floor/wall; refer to Figure-16. If you want to specify dimensions of the floor/wall without dragging the mouse then enter the desired value in **X Dimension** and **Y Dimension** edit boxes in the form.

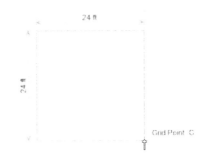

Figure-16. Creating rectangular floor/ wall

Quick Draw Floor/Wall (Plan, Elev) Tool

The **Quick Draw Floor/Wall (Plan, Elev)** tool is used to quickly draw floors/walls in plan and elevation views using the grid blocks. The procedure to use this tool is given next.

- Click on the **Quick Draw Floor/Wall (Plan, Elev)** tool from the **Draw Floor/Wall Objects** cascading menu of the **Draw** menu. You will be asked to select the grid block and the **Properties of Object** form will be displayed.
- Select the desired option from the **Property** drop-down.
- Click at the desired location in the grid block. The floor/wall will be created; refer to Figure-17.

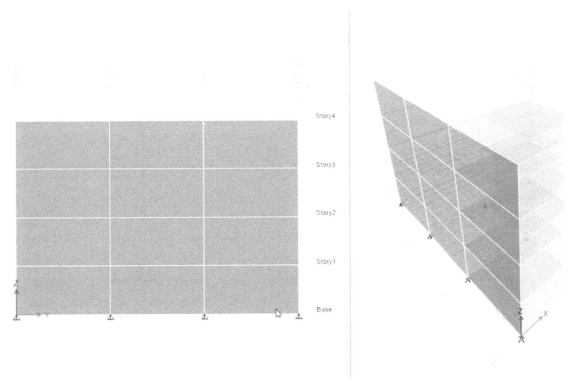

Figure-17. Walls created

- Press **ESC** to exit the tool.

Draw Walls (Plan) Tool

The **Draw Walls (Plan)** tool is used to draw walls in plan view. The procedure to use this tool is given next.

- Click on the **Draw Walls (Plan)** tool from the **Draw Floor/Wall Objects** cascading menu of the **Draw** menu. You will be asked to specify start and end points of the wall. Also, the **Properties of Object** form will be displayed; refer to Figure-18.

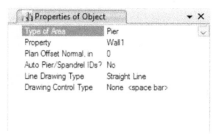

Figure-18. Properties of Object form

- Select the desired option from the **Type of Area** drop-down. There are two options in this drop-down: **Pier** and **Spandrel**. Select the **Pier** option to create the most common type of permanent wall. Select the **Spandrel** option to create exterior curtain walls outside the beams of multistory buildings.
- Set the desired parameters in the fields of **Properties of Object** form and click on the model to specify starting point of the wall. The other end point of the wall will get attached to the cursor and you will be asked to specify the other point.
- Click at the desired location to place the other end point of the wall. The wall will be created; refer to Figure-19 and you will be asked to specify the other end point.

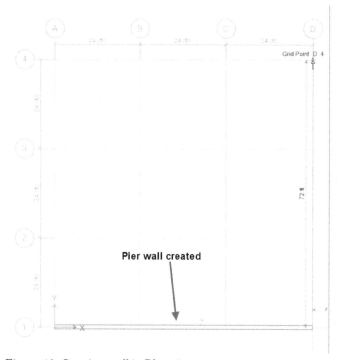

Figure-19. Creating wall in Plan view

- Click to specify the next end point or press ESC to exit.

Similarly, you can use the other tools in the **Draw Floor/Wall Objects** cascading menu.

DRAWING LINKS

Links are used to transfer loads from one point to other. The procedure to draw link is given next.

- Click on the **Draw Links** tool from the **Draw** menu. You will be asked to specify the points to be linked.
- Click at the desired locations to connect them by link; refer to Figure-20.

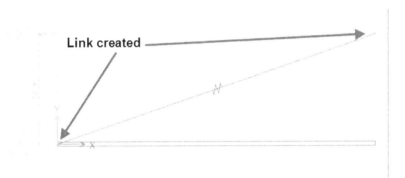

Figure-20. Link created

Drawing Tendons

Tendons are used to simulate the effect of prestressing and post tensioning on the objects like beams, columns, frames etc. The procedure to draw tendons is given next.

- Click on the **Draw Tendons** tool from the **Draw** menu. You will be asked to specify locations for tendons.
- Click at the desired locations to draw tendons. Press ESC to exit the tool.

Drawing Design Strips

Design Strips are used to form regions in a slab where design cuts will automatically generated and flexural reinforcing bars are designed. If the moment varies along the length of design strip then the individual design cuts will show varying moment at each section. The results for the entire Design Strip is governed by the maximum moment demand at an individual design cut. Note that the same number and size of reinforcing bars are used for the entire Design Strip. The procedure to draw design strip is given next.

- Click on the **Draw Design Strips** tool from the **Draw** menu. You will be asked to specify the location points for strip and options related to design strip will be displayed in the **Properties of Object** form.

- Specify the desired parameters in the **Properties of Object** form and click at the desired points to create design strip; refer to Figure-21.

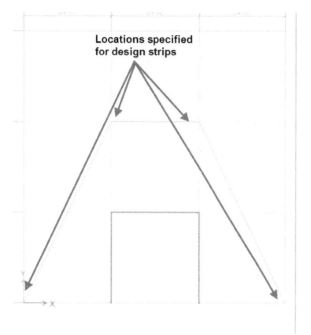

Figure-21. Design strips created

- Press **ESC** to exit the tool.

DRAWING DIMENSION LINES

Dimension lines are used to annotate distance between two points on the drawing or size parameters of any geometric object. The procedure to draw dimension lines is given next.

- Click on the **Draw Dimension Lines** tool from the **Draw** menu. You will be asked to specify the points to measured.
- Click at the desired locations to specify the start and end points of the dimension line. A dashed line will get attached to the cursor; refer to Figure-22.

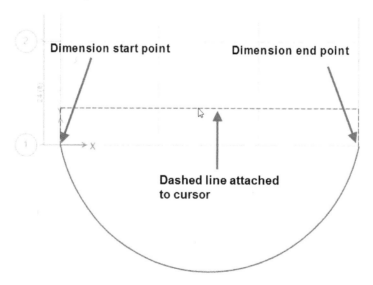

Figure-22. Dashed line for placing dimension

- Click at the desired location to place the dimension.

DRAWING REFERENCE POINTS

Draw Reference Points tool is used to draw reference points. These reference points can be used to create various structural objects in the model. The procedure to use this tool is given next.

- Click on the **Draw Reference Points** tool from the **Draw** menu. You will be asked to specify the location of reference point and the **Properties of Object** form will be displayed.
- Specify the desired offset values in the X, Y, and Z direction if required. Select the desired grid system from the **Grid System** drop-down in the form.
- Click at the desired locations in the viewport to create the reference points.

SETTING VIEWS IN VIEWPORTS

Using the tools in toolbar, you can set different views in the viewports as required. The procedures to setting different views are given next.

Setting 3D View

- Click in the desired viewport while no tool is active.
- Click on **Set Default 3D View** tool ₃-d from the toolbar to set the current viewport to 3D view.

Setting Plan View

- Click on the **Set Plan View** tool pla n from the toolbar to set the current viewport to plan view. The **Select Plan View** dialog box will be displayed; refer to Figure-23.

Figure-23. Select Plan View dialog box

- Select the desired story from the list to set it in plan view as working plane.

- If you want to set a user defined height as working plane location then select the **At Specified Global Z Coordinate** radio button from the top area of the dialog box. The **Global Z Coordinate** edit box will be displayed below it.
- Specify the desired height value in the edit box using the Global coordinate system.
- After setting the desired options, click on the **OK** button.

Setting Elevation View

- Click on the **Set Elevation View** tool ᵉˡᵉᵥ from the toolbar after clicking in the viewport whose view is to be changed. The **Set Elevation View** dialog box will be displayed; refer to Figure-24.

Figure-24. Set Elevation View dialog box

- Select the desired option from the list.
- If you want to create elevation view at desired X or Y ordinate then click on the **Add at Ordinate** button from the dialog box. The **Add Elevation View At Ordinate** dialog box will be displayed; refer to Figure-25.

Figure-25. Add Elevation View At Ordinate dialog box

- Select the desired radio button and specify the ordinate value. Click on the **OK** button to add the ordinate to current list of elevations.
- Click on the **OK** button from the **Set Elevation View** dialog box to set the current viewport to elevation view.

DRAWING REFERENCE PLANES

The reference planes are used to provide horizontal planes/lines that can be used to snap to when drawing objects in elevation views. The procedure to draw reference plane is given next.

- Set a viewport to desired elevation view as discussed earlier.
- Click on the **Draw Reference Planes** tool from **Draw** menu. You will be asked to specify location for reference plane.
- Set the desired offset value in the **Vertical Offset Z** edit box of the **Properties of Object** form.
- Click at the desired location to place the reference. The reference plane will be created; refer to Figure-26.

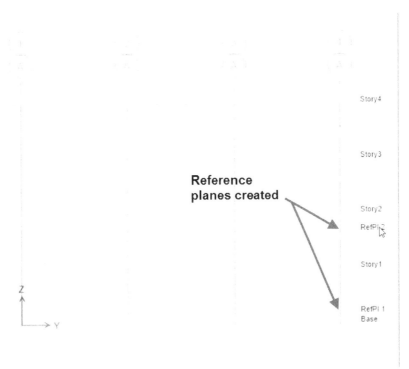

Figure-26. Reference plane created

DRAW DEVELOPED ELEVATION DEFINITION

The developed elevation is used to display selected elevation in unfolded position. The procedure to draw developed elevation is given next.

- Make sure to have plan view in viewport. Click on the **Draw Developed Elevation Definition** tool from the **Draw** menu. The **Developed Elevation Name** dialog box will be displayed; refer to Figure-27.

Figure-27. Developed Elevation Name
dialog box

- Specify the desired name in the edit box and click on the **OK** button. You will be asked to create a multiline over all the walls/beams/ objects to be included in the developed elevation.
- Create a multiline as required; refer to Figure-28.

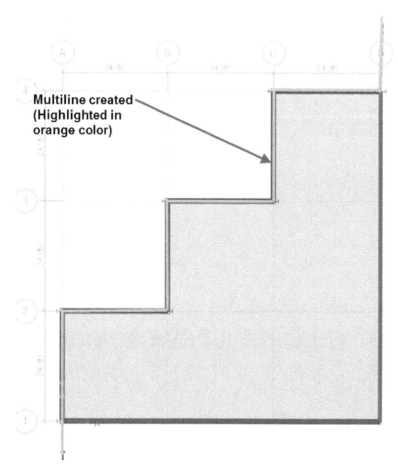

Figure-28. Multiline created for developed elevation

- Press **ENTER** after drawing the multiline for developed elevation. The developed elevation will be displayed in the viewport; refer to Figure-29.

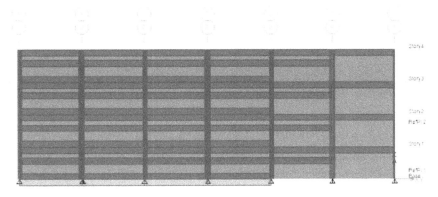

Figure-29. Developed elevation created

DRAWING WALL STACK

Wall stack is a complex combination of walls. Make sure plan view is active in the viewport as it makes easy to place wall stacks in the building. The procedure to draw wall stack in ETABS is given next.

- Set the viewport to plan view. Click on the **Draw Wall Stacks (Plan, Elev, 3D)** tool from the **Draw** menu. The **New Wall Stack** dialog box will be displayed; refer to Figure-30.

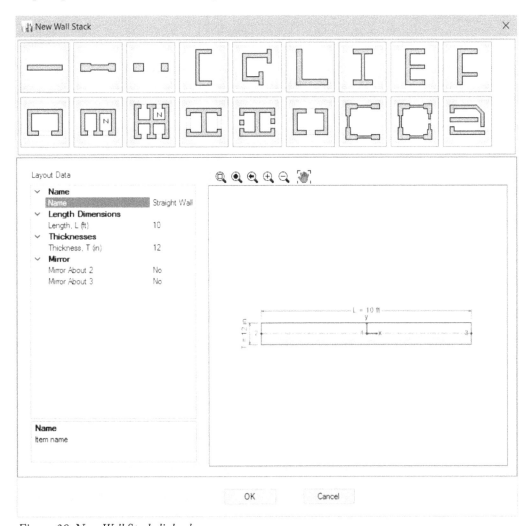

Figure-30. New Wall Stack dialog box

- Select the desired button from the palette to select a shape of walls. The parameters for selected shape will be displayed along with preview in the dialog box.
- Set the desired parameters in the dialog box and click on the **OK** button. The select wall shape will get attached to cursor and you will be asked to specify the insertions point for the wall stack.
- Click at the desired locations in the plan view to place wall stacks; refer to Figure-31.

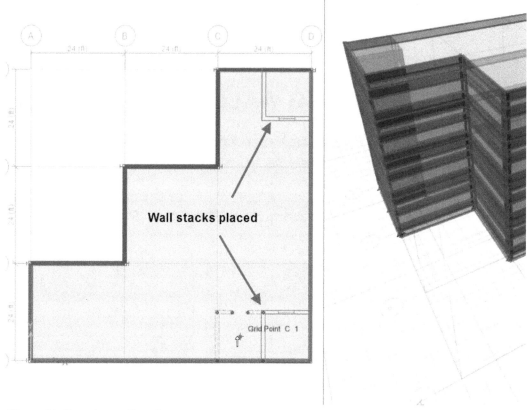

Figure-31. Drawing wall stacks

- Note that you can change the orientation of wall stack by specifying desired angle value in the **Angle, deg** edit box of the **Properties of Object** form. To define the limit of wall stacks vertically, select the desired options in the **Top Story** and **Bottom Story** drop-downs.
- Press ESC to exit the tool.

AUTO DRAW CLADDING

The **Auto Draw Cladding** tool in the **Draw** menu is used to draw cladding based on specified parameters automatically. Cladding is used to provide thermal insulation or weather protection. The procedure to use the **Auto Draw Cladding** tool is given next.

- Click on the **Auto Draw Cladding** tool from the **Draw** menu. The **Cladding Options** dialog box will be displayed; refer to Figure-32.

Figure-32. Cladding Options dialog box

- Select the desired option from the dialog box and click on the **OK** button (We have used beams to create cladding so **Use Beams** radio button is selected). The cladding will be created automatically.

PRACTICAL

Create the structure as shown in Figure-33.

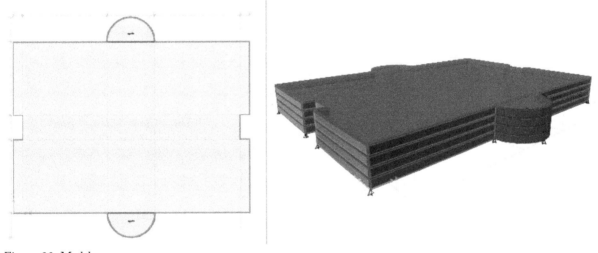

Figure-33. Model

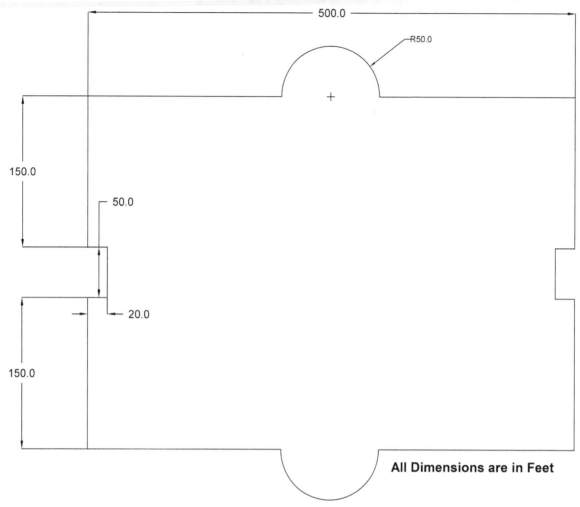

Figure-34. Dimensions for model

Steps:

Creating Grid Data

- Start ETABS if not started yet.
- Click on the **New** button from the toolbar. The **Model Initialization** dialog box will be displayed.
- Select the **Use Built-in Settings With:** radio button and set the parameters as U.S. Customary, AISC14, AISC 360-10, and ACI 318-14 in the drop-downs. Click on the **OK** button. The **New Model Quick Templates** dialog box will be displayed.
- Select the **Custom Grid Spacing** radio button and click on the **Edit Grid Data** button. The **Grid System Data** dialog box will be displayed; refer to Figure-35.

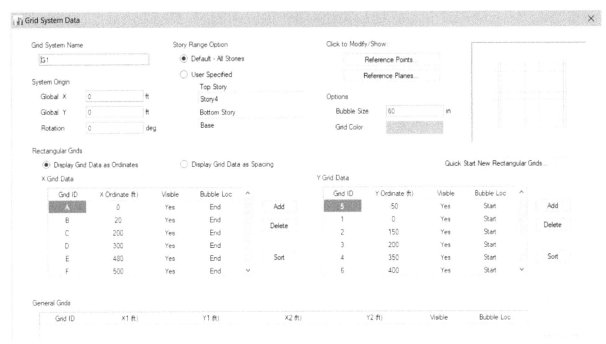

Figure-35. Grid System Data dialog box

- Set the name as Commercial in the **Grid System Name** edit box.
- Set the grid points in X Grid Data table as **0, 20, 200, 300, 480, and 500**; refer to Figure-35.
- Set the grid points in Y Grid Data table as **-50, 0, 150, 200, 350, and 400**; refer to Figure-35.
- After setting desired parameters, click on the **OK** button. The **New Model Quick Templates** dialog box will be displayed again.
- Click on the **OK** button from the dialog box. The grids will be displayed in the plan and 3D view; refer to Figure-36.

Figure-36. Grids displayed in viewport

Drawing Beam and Editing Point of Beams

- Click on the **Draw Beam/Column/Brace (Plan, Elev, 3D)** tool from the **Draw Beam/Column/Brace Objects** cascading menu of **Draw** menu. You will be asked to specify starting point of the beam.
- Select the **All Stories** option from the **Stories** drop-down and create the beam frame as shown in Figure-37.

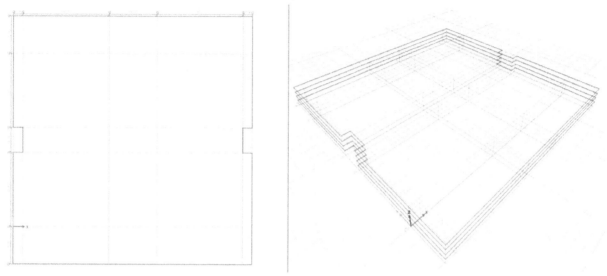

Figure-37. Creating beam structure

- Delete the outer beams and draw new beams as shown in Figure-38.

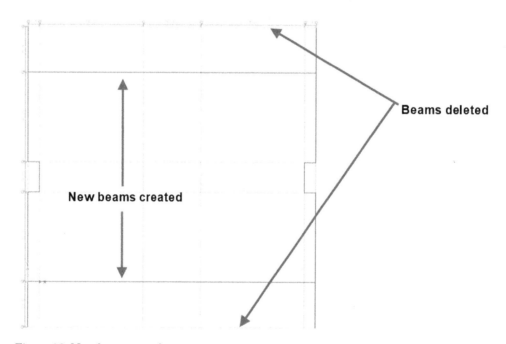

Figure-38. New beams created

- Now, click on the **Reshape Object** tool from the **Draw** menu. You will be asked to select objects to be reshaped. Select the beam as shown in Figure-39.

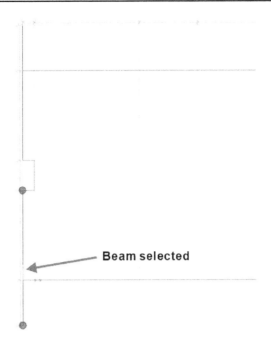

Figure-39. Beam selected

- Drag bottom end point of the beam selected and link it to the newly created beam; refer to Figure-40.

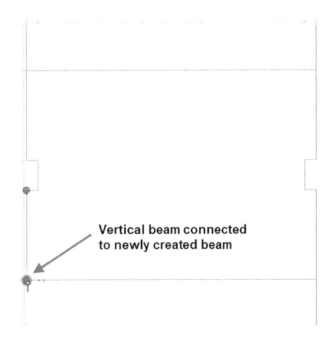

Figure-40. Dragging point to connect to beam

- Set the other extending beams in the same way; refer to Figure-41.

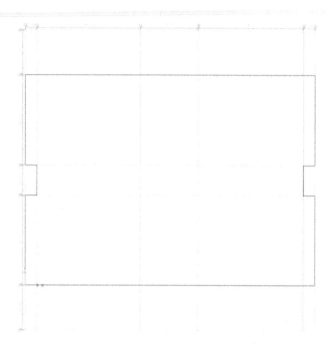

Figure-41. Reshaping beams

- Click on the **Draw Beam/Column/Brace (Plan, Elev, 3D)** tool from the **Draw Beam/Column/Brace Objects** cascading menu of **Draw** menu. You will be asked to specify points of beam.
- Select the **Arc (3 Points)** option from the **Line Drawing Type** drop-down of **Properties of Object** tab and create arc beams as shown in Figure-42.

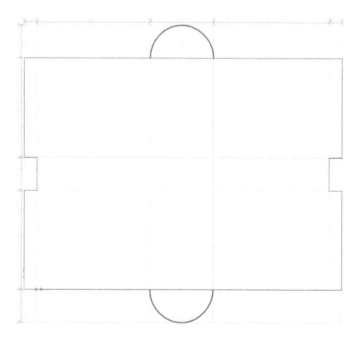

Figure-42. Arc beams created

Drawing Floors

- Click on the **Draw Floor/Wall (Plan, Elev, 3D)** tool from the **Draw Floor/ Wall Objects** cascading menu. You will be asked to specify start point of the floor.
- Draw the floor as shown in Figure-43. Press ESC to exit the tool.

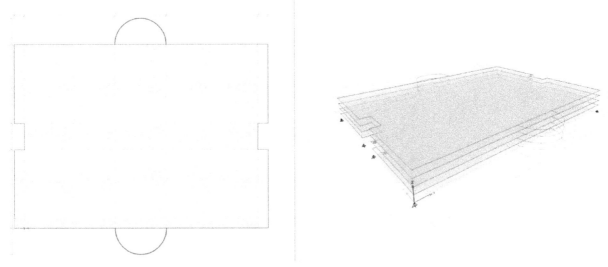

Figure-43. Creating floor slab

- Activate the tool again and select the **Deck 1** option from the **Property** drop-down in the **Properties of Object** tab.
- Select the **Arc (3 Points)** option from the **Edge Drawing Type** drop-down in **Properties of Object** tab and draw the deck in arc beams; refer to Figure-44.

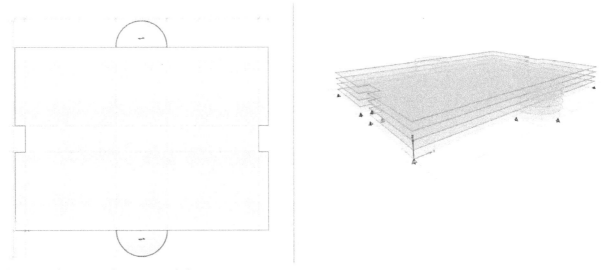

Figure-44. Structure after creating deck

Creating Columns

- Click on the **Quick Draw Columns (Plan, 3D)** tool from the **Draw Beam/ Column/Brace Objects** cascading menu of the **Draw** menu. You will be asked to select points for creating columns.
- Click at the grid points encircled in Figure-45. The columns will be created; refer to Figure-46.

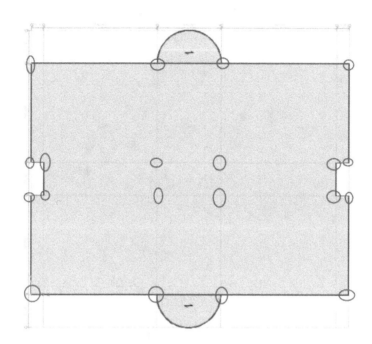

Figure-45. Points to be selected for column

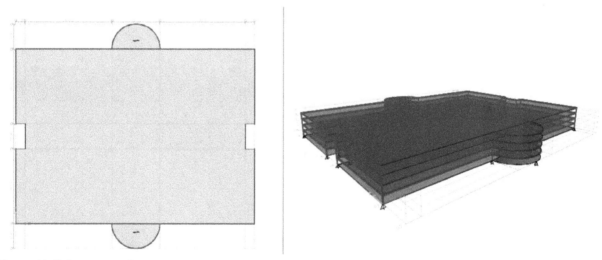

Figure-46. Columns created

Creating Walls

- Click on the **Draw Walls (Plan)** tool from the **Draw Floor/Wall Objects** cascading menu of the **Draw** menu. You will be asked to select start point for wall.
- Select the **Straight Line** option from the **Line Drawing Type** drop-down in **Properties of Object** tab and create the wall as shown in Figure-47.

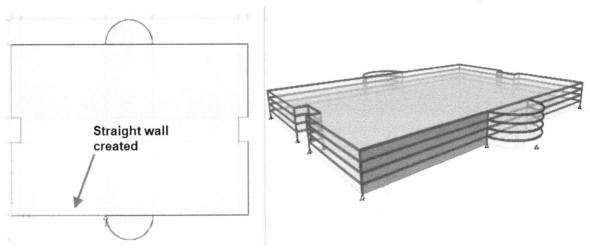

Figure-47. Straight wall created

- Select the **Arc (3 Points)** option from the **Line Drawing Type** drop-down in the **Properties of Object** tab and create semicircular wall as shown in Figure-48.

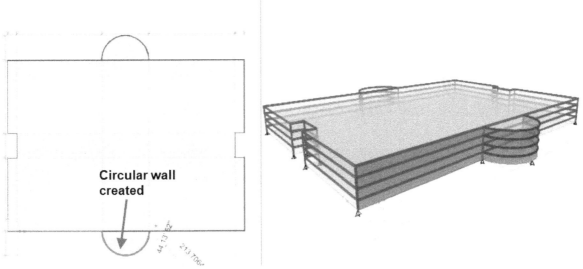

Figure-48. Circular wall created

- Select the option from the **Line Drawing Type** drop-down as required and create rest of the walls as shown in Figure-49. Press **ESC** to exit the tool.

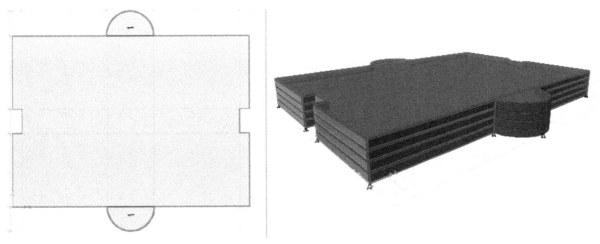

Figure-49. Creating walls

- Save the model and close the file.

FOR STUDENT NOTES

Chapter 3

Section Properties and Material

Topics Covered

The major topics covered in this chapter are:

- *Introduction*
- *Defining Section Properties*
- *Section Designer*
- *Drawing Shapes*
- *Defining Tendon Sections*
- *Defining Slab Sections*
- *Defining Deck Sections*
- *Defining Wall Sections*

- *Defining Reinforcing Bar Sizes*
- *Defining Link/Support Properties*
- *Defining Frame/Wall Hinge Properties*
- *Defining Panel Zone*
- *Defining Spring Properties*
- *Diaphragm Properties*
- *Pier Labels and Spandrel Labels*
- *Creating Groups*

INTRODUCTION

In the previous chapter, you have learned to draw different objects in ETABS. But we have not discussed about the type of sections used in drawing these objects. Like, we have not discussed the section type being used to draw beam/column. We have also not discussed about the materials and their properties for objects. In this chapter, we will discussed all these topics.

DEFINING SECTION PROPERTIES

The section properties are the shape and material definitions of building blocks used to draw objects in ETABS. For examples, beams used in building have some section properties, slabs used to make floor have some section properties, and so on. The tools to define section properties are available in the **Section Properties** cascading menu of the **Define** menu; refer to Figure-1. The tools in this menu are discussed next.

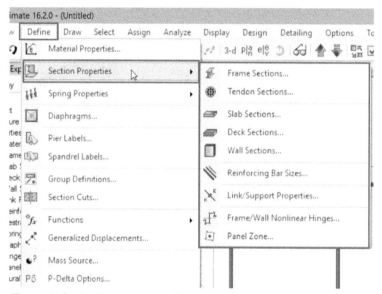

Figure-1. Section Properties cascading menu

Defining Frame Sections

The **Frame Section** tool in the **Section Properties** cascading menu of the **Define** menu is used to define the shape, size, and material of different frame sections. The procedure to define frame section is given next.

• Click on the **Frame Section** tool from the **Section Properties** cascading menu of the **Define** menu. The **Frame Properties** dialog box will be displayed; refer to Figure-2.

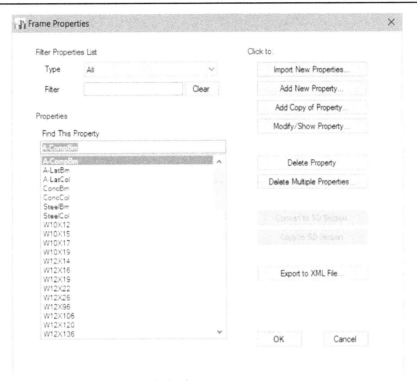

Figure-2. Frame Properties dialog box

- By default, all the frame properties are displayed in the **Properties** area of the dialog box.

Importing New Properties

- Click on the **Import New Properties** button from the dialog box. The **Frame Property Shape Type** dialog box will be displayed; refer to Figure-3.

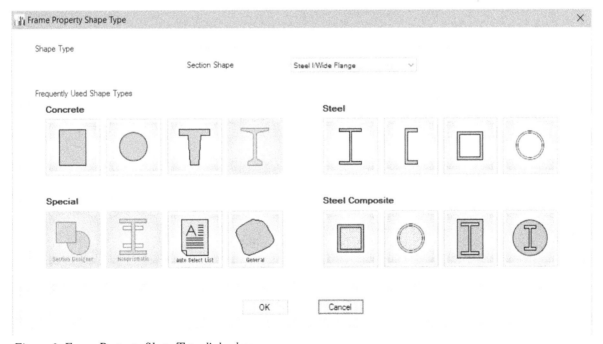

Figure-3. Frame Property Shape Type dialog box

- Select the desired shape type from the **Section Shape** drop-down and click on the **OK** button or select the desired button from the **Frequently Used Shape Types** area of the dialog box. The **Frame Section Property Import Data** dialog box will be displayed; refer to Figure-4.

Figure-4. Frame Section Property Import Data dialog box

- Select the desired XML property file from the **Name of XML Property File** drop-down. If you have an XML property file saved in your local drive then click on the Browse button next to the drop-down. The **Add XML Property Files** dialog box will be displayed; refer to Figure-5. Select the desired file and click on the **Open** button. The options related to selected property will be displayed in the dialog box.

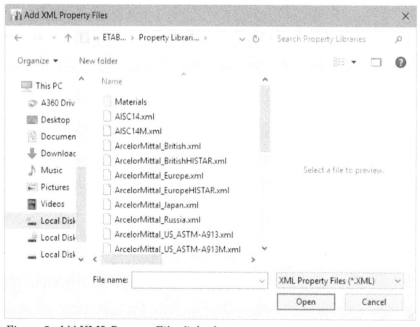

Figure-5. Add XML Property Files dialog box

- Select the desired material from the **Default Material for Section** drop-down.
- Change the option in **Section Shape Type** drop-down as required. The list of section properties that can be imported, will be displayed in the **Select Section Properties to Import** area of the dialog box.
- Select the section from the list and click on the **OK** button from the dialog box. The section will be added in the list.

Adding New Section Property

The **Add New Section Property** button in the **Frame Properties** dialog box is used to add new section properties in the list. The procedure is given next.

- Click on the **Add New Section Property** button from the dialog box. The **Frame Property Shape Type** dialog box will be displayed as discussed earlier.
- Select the desired shape type from the **Section Shape** drop-down and click on the **OK** button or select the desired button from the **Frequently Used Shape Types** area of the dialog box. The **Frame Section Property Import Data** dialog box will be displayed; refer to Figure-6.

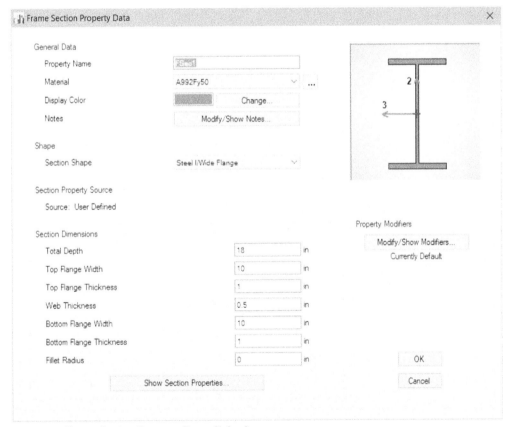

Figure-6. Frame Section Property Data dialog box

- Specify desired name of the section in **Property Name** edit box.
- Select the desired material from the **Material** drop-down.
- Click on the **Change** button for **Display Color** option to change the color of the frame section.
- You can add a note related to the current section by using the **Modify/Show Note** button in the **General Data** area of the dialog box.

On clicking this button, the **Frame Property Notes** dialog box is displayed. Specify the desired text note and click on the **OK** button to apply changes.

• If you want to change the section shape from what you have selected earlier, click in the **Section Shape** drop-down of the **Shape** area in the dialog box and select the desired option. Note that based on your selection in this drop-down, the options of the **Section Dimensions** area get changed.

• Set the section dimensions as required in the dialog box.

• Click on the **Modify/Show Modifiers** option to check/change the parameters directly related to structural analysis. The **Property/ Stiffness Modification Factors** dialog box will be displayed; refer to Figure-7. Set the multiplication factors for desired properties to modify it and click on the **OK** button.

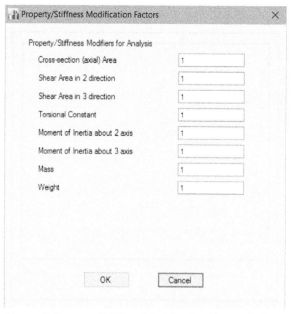

Figure-7. Property Stiffness Modification Factors dialog box

• Click on the **OK** button from **Frame Section Property Data** dialog box. The new property will be added in the list.

Adding Copy of a Property

• Select the property from the list whose copy is to be added and click on the **Add Copy of Property** button. The **Frame Section Property Data** dialog box will be displayed with all the parameters of selected property.

• Note that if you have selected a standard property then **Section Dimensions** area of the dialog box will not be active; refer to Figure-8. To make this area active for modifications, click on the

Convert to User Defined button.

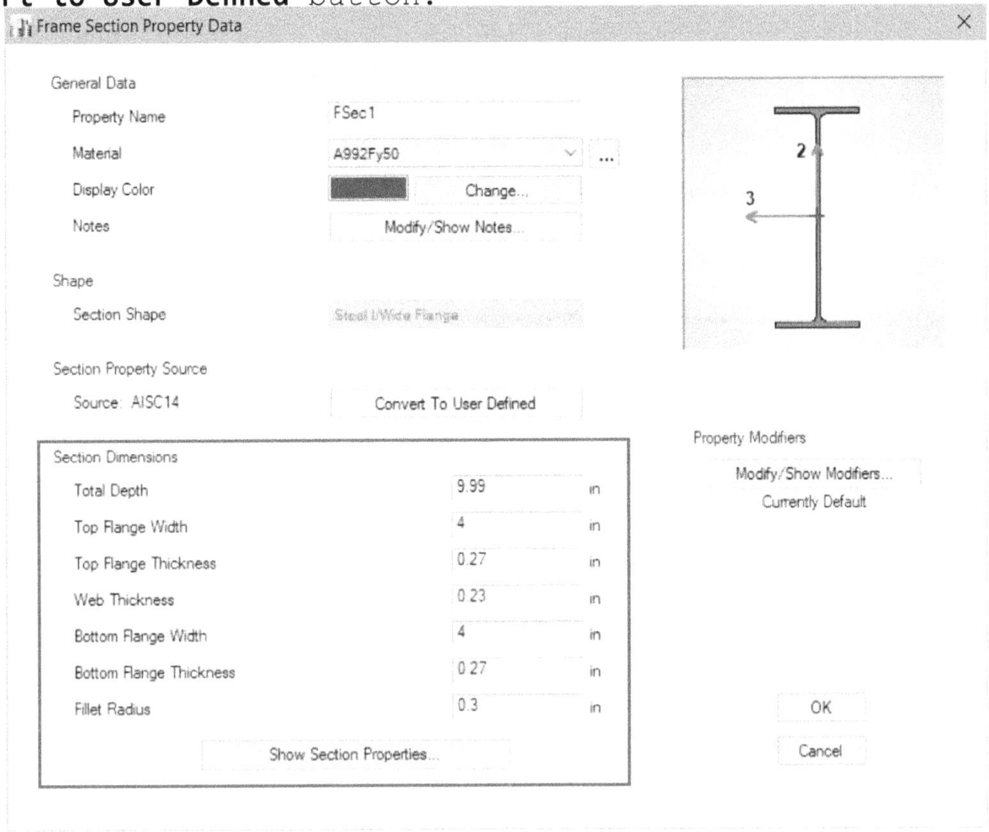

Figure-8. Section Dimensions area of Frame Section Property Data dialog box

- Perform the modifications as required and click on the **OK** button from the dialog box to add the property in the list.

Modifying Property Data

- To modify a section property, select it from the list and click on the **Modify/Show Property** button from the **Frame Properties** dialog box. The **Frame Section Property Data** dialog box will be displayed as discussed earlier.
- Set the desired parameters and click on the **OK** button from the dialog box.

Deleting a Property

Select the section property from the list that you want to be deleted and click on the **Delete Property** button from the **Frame Properties** dialog box. A confirmation box will be displayed. Click on the **Yes** button to delete.

Deleting Multiple Properties

The **Delete Multiple Properties** button is used to delete all the properties that can be deleted. The procedure is given next.

- Click on the **Delete Multiple Properties** button from the **Frame Properties** dialog box. The **Delete Multiple Frame Section Properties** dialog box will be displayed; refer to Figure-9.

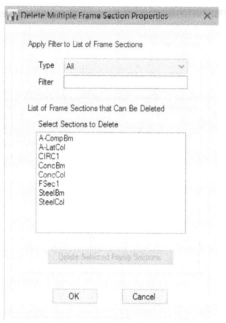

Figure-9. Delete Multiple Frame Section
Properties dialog box

- Select the frame sections that you want to be deleted from the list while holding the **CTRL** key and click on the **Delete Selected Frame Sections** button. The selected sections will get deleted.
- Click on the **OK** button to exit the dialog box.

Converting a Standard Section to SD Section

SD stands for Section Designer. SD sections are the sections of frame member with irregular shape manually drawn by the user. So, the benefit of converting a frame member to section designer section is that you can now change its shape as required. You should always make sure that your designed section is available in market or can be produced up to your required quantity. The procedure to convert a standard section to SD section is given next.

- Select the section from the list in the **Frame Properties** dialog box and click on the **Convert to SD Section** button. The **Section Designer Section Property Data** dialog box will be displayed; refer to Figure-10.
- Specify the name as desired in the **Property Name** edit box. Change the other parameters as required in the **General Data** area of the dialog box.
- In the **Design Type** area, select the desired radio button. Note that radio buttons in this area are active based on the material selected in **Base Material** drop-down. On the basis of radio button selected from the **Design Type** area, the system will analyze the model using the respective post processor. For example, if you have selected the **Composite Column** radio button then system will analyze model using the **Composite Column Post Processor**.
- Selecting the **Concrete Column** or **Composite Column** radio button will activate the radio buttons available in the **Concrete Column Check/ Design** area of the dialog box.

Figure-10. Section Designer Section Property Data dialog box

- Select the desired option from the **Concrete Column Check/Design** area so that **Concrete Frame Design Post Processor** checks or designs the reinforcement in the column.
- Click on the **Section Designer** button from the **Define/Edit/Show Section** area to check or modify the section. The **Section Designer** window will be displayed; refer to Figure-11.

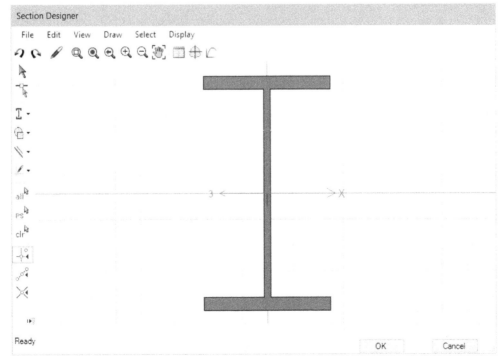

Figure-11. Section Designer window

- Create/modify the shape of section as required and click on the **OK** button from the window. The **Section Designer Section Property Data** dialog box will be displayed again. You will learn more about the tools in **Section Designer** window later in this book.
- Set the other properties as discussed earlier and click on the **OK** button. The modified section will be displayed in the list.

Similarly, you can use the **Copy to SD Section** button to create a copy of selected section property as SD section.

Exporting to XML File

The **Export to XML File** tool in the **Frame Properties** dialog box is used to export the selected section property to an XML file. The procedure is given next.

- After selecting the desired property, click on the **Export to XML File** button from the dialog box. The **Frame Property Export to XML File** dialog box will be displayed; refer to Figure-12.

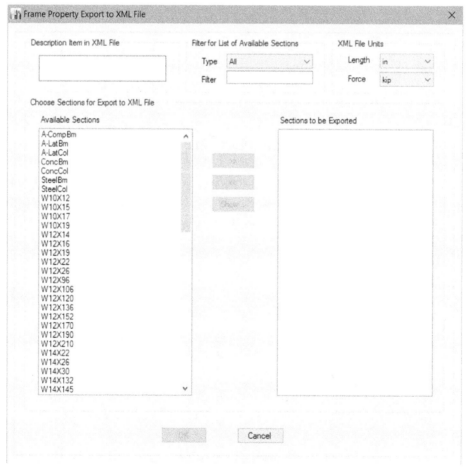

Figure-12. Frame Property Export to XML File dialog box

- Select all the properties that you want to be exported to an xml file from the left list box while holding the **CTRL** key.
- Click on the **>>** button to add selected files to export list.

- Set the unit of Length and Force in the respective drop-downs of **XML File Units** area of the dialog box.
- Click on the **OK** button after setting the parameters. The **Save Frame Property XML File As** dialog box will be displayed; refer to Figure-13.

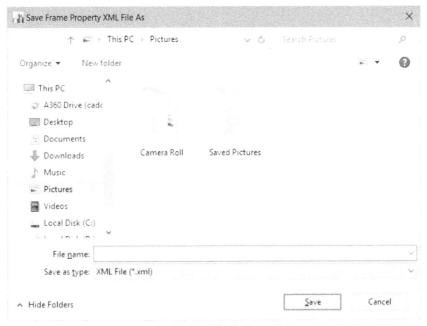

Figure-13. Save Frame Property XML File As dialog box

- Specify the name of file and save it in the desired location by clicking **Save** button. A confirmation box will be displayed.
- Click on the **OK** button to exit the confirmation box.

Click on the **OK** button from the **Frame Properties** dialog box after setting the parameters as required. Note that all the new or modified properties set in **Frame Properties** dialog box can now be accessed while creating beams/columns/braces using the tools discussed in previous chapter.

SECTION DESIGNER

The **Section Designer** window is used to draw irregular designs or combine two or more standard designs of frame sections to create a new frame section. The **Section Designer** window is displayed on clicking the **Section Designer** button from the **Define/Edit/Show Section** area of the **Section Designer Section Property Data** dialog box. We have discussed about activating the **Section Designer** window in previous topic. Now, we will discussed the procedure to create different sections in **Section Designer** window. Most of the tools in this window are same as available the ETABS interface. We will discussed the tools used for drawing sections in this topic.

Drawing a section involve many steps in Section Designer like activating required snaps, creating reference geometries, drawing different shapes, selecting and reshaping the objects, and checking different properties. These processes are discussed next.

Activating Snaps

There are six options to activate different type of snaps. These options are available in **Snap To** cascading menu of **Draw** menu in the window and at the bottom of the toolbar in the left of window; refer to Figure-14.

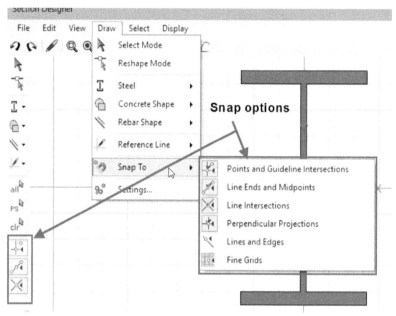

Figure-14. Snapping options

- Click on the desired snap option from the **Snap To** cascading menu or select the button from the left toolbar. The respective snap option will be activated.
- To de-activate a snap, click on the respective option again.

Select the **Point and Guideline intersection** option to snap cursor to the points in the object and points generated by entities at their intersection.

Select the **Line Ends and Midpoint** option to snap cursor to ends points and mid points of the lines in the object.

Select the **Line Intersections** option to snap cursor to the intersection points of the lines.

Select the **Perpendicular Projections** option to snap cursor to the perpendicular project of current geometry being created; refer to Figure-15.

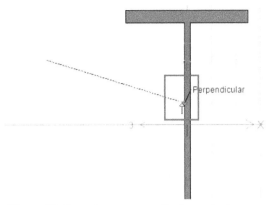

Figure-15. Snapping to perpendicular projection

Select the **Lines and Edges** option to snap cursor to lines and edges of the other objects in the section.

Select the **Fine Grids** option to snap cursor to grid lines.

Creating Reference Geometries

The tools to create reference geometries are available in the **Reference Line** cascading menu of the **Draw** menu and **Draw Reference Shape** toolbox; refer to Figure-16. The procedures to draw different reference shapes are given next.

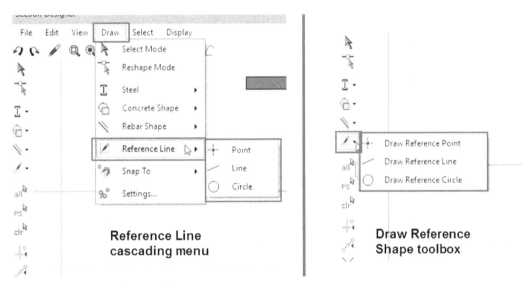

Figure-16. Draw Reference tools

Drawing Reference Point

- Click on the **Draw Reference Point** tool from the **Reference Line** cascading menu of the **Draw** menu. You will be asked to left click at desired location to specify position of reference point.
- Click at the desired location to place the reference point. The reference point will be created.

Drawing Reference Line

- Click on the **Draw Reference Line** tool from the **Reference Line** cascading menu of the **Draw** menu. You will be asked to specify the start point of the line.
- Click at the desired location. A rubber band line will get attached to the cursor and you will be asked to specify the end point of the reference line.
- Click at the desired location to draw the line.

Drawing Reference Circle

- Click on the **Draw Reference Circle** tool from the **Reference Line** cascading menu of the **Draw** menu. You will be asked to specify the center point of the circle.
- Click at the desired location. The circle will be drawn. (What? You have not given radius/diameter!!)
- Note that the system automatically takes default diameter viz. 24 inches. To change the diameter, click on the **Select Object** tool from the left toolbar and select the circle. Now, right-click on the circle. The **Section Object Data - ReferenceCircle** dialog box will be displayed; refer to Figure-17. Note that a similar dialog box is displayed when you right-click on any object in the ETABS model. Using the options in this dialog box, you can change the parameters of the object.

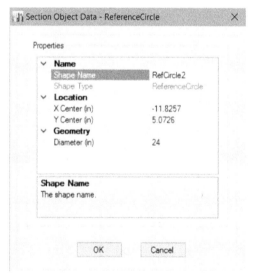

Figure-17. Section Object Data-ReferenceCircle dialog box

- Click in the **Diameter (in)** edit box of **Geometry** node and specify the desired value of diameter.
- You can also change the other properties of reference circle. Click on the **OK** button from the dialog box to apply the changes.

Drawing Shapes

The tools to draw shapes are categorized in three sections: Steel Shapes, Concrete Shapes, and Rebar Shapes. In Section Designer, these tools are available in similarly named three cascading menus viz. **Steel, Concrete Shape,** and **Rebar Shape**; refer to Figure-18, Figure-19, and Figure-20.

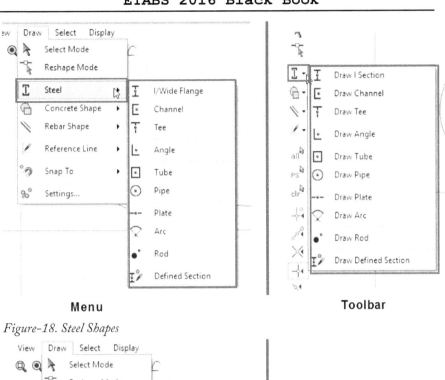

Menu   **Toolbar**

Figure-18. Steel Shapes

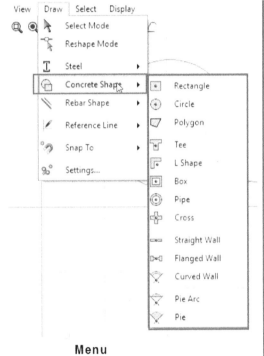

Menu   **Toolbar**

Figure-19. Concreate Shape tools

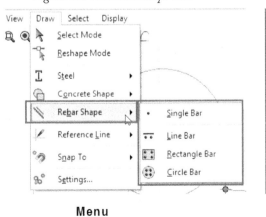

Menu

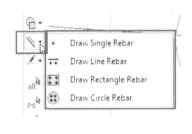

Toolbar

Figure-20. Rebar Shape tools

Drawing Shapes

- Click on the **Draw Channel** tool from the **Steel** cascading menu of the **Draw** menu or click on the **Draw Channel** tool from the **Draw Steel Shape** toolbox. You will be asked to specify the center location of steel channel.
- Click at the desired location to place the channel. The channel will be drawn at specified location; refer to Figure-21. Select the channel and right-click on it to change its parameters.

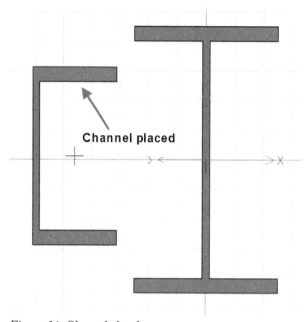

Figure-21. Channel placed

Similarly, you can draw other steel, concrete, and rebar shapes.

Drawing Defined Section

The **Draw Defined Section** tool in the **Steel** cascading menu of **Draw** menu is used to insert steel shapes already available in the library. The procedure to use this tool is given next.

- Click on the **Draw Defined Section** tool from the **Steel** cascading menu. You will be asked to specify the placement location.
- Click at the desired location. The **Defined Steel Section** dialog box will be displayed; refer to Figure-22.

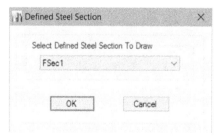

Figure-22. Defined Steel Section dialog box

- Select the desired option from the drop-down and click on the **OK** button. The section will be placed.

Reshaping the Objects

The **Reshape Mode** option is used to activate the mode in which you can drag the points of the object to reshape it. The procedure is given next.

- Click on the **Reshape Mode** tool from the **Draw** menu. You will be asked to select a point on the model.
- Click at the desired location on the object. Nodes will be displayed on the object to modify its shape; refer to Figure-23.

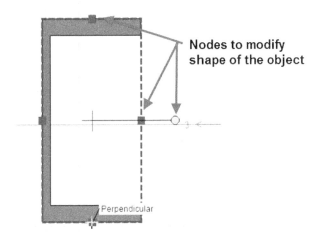

Figure-23. Nodes to modify shape of object

- Drag the nodes to modify the shape of object; refer to Figure-24.

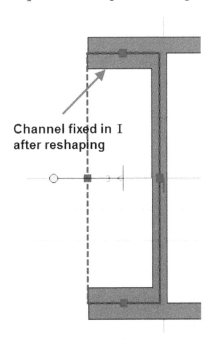

Figure-24. Reshaping objects

Checking Properties of Section

Once you have created the section, the next step is to check its properties. There are three tools in the **Display** menu; refer to Figure-25.

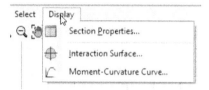

Figure-25. Display menu

The procedures to check different properties are given next.

Checking Section Properties

Section Properties are properties like center of gravity location, section modulus, moment of inertia, and so on. The procedure to check these properties for designed section is given next.

• Click on the **Section Properties** tool from the **Display** menu. The **Section Properties** dialog box will be displayed; refer to Figure-26.

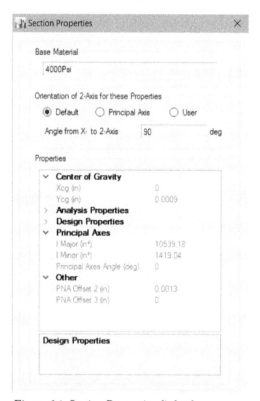

Figure-26. Section Properties dialog box

• Select the desired radio button from the **Orientation of 2-Axis for these Properties** area to define orientation for checking properties.

Check the properties from the **Properties** area. Click on the **X** button at the top right corner to close the dialog box. Note that reinforced bars are not considered in section properties and following section properties are displayed in the **Properties** area of the dialog box.

X_{cg} and Y_{cg}	Location of Center of Gravity.
A	Area of Section.
AS2	Shear area for shear parallel to the 2-axis.
AS3	Shear area for shear parallel to the 3-axis.
I22 and I33	Moment of Inertia about 2-axis and 3-axis, respectively
I23	Moment of Inertia when axis 2 and 3 are not principle axis which is never the case in ETABS. So, the value of I23 is always zero.
J	Torsional Constant.
R22 and R33	Radius of Gyration about 2-axis and 3-axis.
S22(+), S22(-)	Section modulus about the 2-axis at extreme fiber of the section in the positive and negative 3-axis direction, respectively.
S33(+), S33(-)	Section modulus about the 3-axis at extreme fiber of the section in the positive and negative 2-axis direction, respectively.
Z22 and Z33	Plastic Modulus of bending for 2-axis and 3-axis respectively.
I Major	Moment of Inertia about major principle axis.
I Minor	Moment of Inertia about minor principle axis.
PNA Offset 2	Offset distance from Center of Gravity along 2-axis.
PNA Offset 3	Offset distance from Center of Gravity along 3-axis.

Checking the Interaction Surface

The **Interaction Surface** tool in the **Display** menu is used to check the interaction between reinforcement and composite material around it in concrete sections. The procedure is given next.

- Click on the **Interaction Surface** tool from the **Display** menu. The **Interaction Surface (ACI 318-14)** dialog box will be displayed; refer to Figure-27.

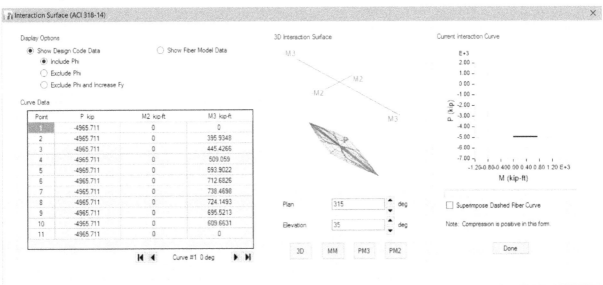

Figure-27. Interaction Surface dialog box

- Check the desired modes of the interaction by selecting the button from the **3D**, **MM**, **PM3**, and **PM2** buttons.
- After checking the properties, click on the **Done** button.

Checking Moment Curvature Plot

The moment curvature plot is used to check the capability of reinforced concrete to bear load up to specified curvature. The procedure to check the plot is given next.

- Click on the **Moment-Curvature Curve** tool from the **Display** menu. The **Moment Curvature Plot** dialog box will be displayed; refer to Figure-28.

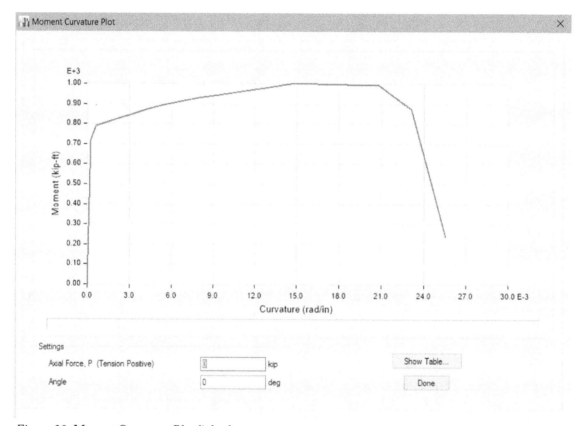

Figure-28. Moment Curvature Plot dialog box

- The vertical scale shows moment and horizontal scale shows curvature of the concrete section. Set the desired values in **Axial Force** and **Angle** edit boxes in the **Settings** area to check plot in different loading conditions.
- Click on the **Show Table** button to check the data in table form.
- After checking the parameters, click on the **Done** button to exit dialog box.

Once you are satisfied with the section design, click on the **OK** button from the **Section Designer** window.

DEFINING TENDON SECTIONS

The tendons are used to simulate effect of pre-stressing and post-tensioning on the model. The procedure to define tendon section is given next.

- Click on the **Tendon Sections** tool from the **Section Properties** cascading menu of the **Define** menu. The **Tendon Properties** dialog box will be displayed; refer to Figure-29.

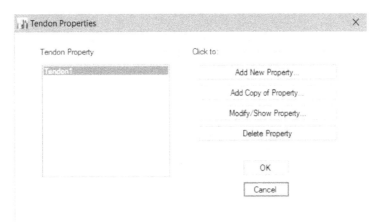

Figure-29. Tendon Properties dialog box

Using the options in this dialog box, you can add a new property, modify an existing property, delete a property and so on. The procedures are given next.

Adding a New Tendon Property

- Click on the **Add New Property** button from the dialog box. The Tendon Property Data dialog box will be displayed; refer to Figure-30.

Figure-30. Tendon Property Data dialog box

- Specify the desired name for the tendon property in **Property Name** edit box.
- Click on the Browse button for **Material Type** and select the desired material.
- Specify the desired strand area value in **Strand Area** edit box and set the other options as required.
- Click on the **OK** button from the dialog box to create the new property.

Adding Copy of a Tendon Property

Click on the **Add Copy of Property** button after selecting the property that you want to be copied from the list. The **Tendon Property Data** dialog box will be displayed as discussed earlier with copied property data. Set the desired parameters and click on the **OK** button to create the new property.

Modifying a Tendon Property

To modify an existing property, select it from the list and click on the **Modify/Show Property** button from the dialog box. The **Tendon Property Data** dialog box will be displayed as discussed earlier. Set the desired parameters and click on the **OK** button to modify the property.

Deleting a Tendon Property

To delete a tendon, select it from the list and click on the **Delete Property** button.

DEFINING SLAB SECTIONS

Slabs are the reinforced concrete blocks used to construct floors or ceilings. The procedure to define sections for slabs is given next.

- Click on the **Slab Sections** tool from the **Section Properties** cascading menu in the **Define** menu. The **Slab Properties** dialog box will be displayed; refer to Figure-31.

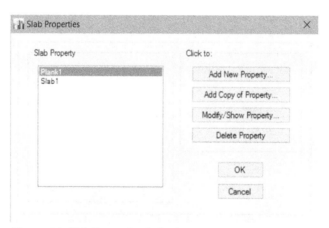

Figure-31. Slab Properties dialog box

- Click on the **Add New Property** button from the dialog box to create a new slab property. The **Slab Property Data** dialog box will be displayed; refer to Figure-32.

Figure-32. Slab Property Data dialog box

- Click in the **Property Name** edit box and specify the desired name.
- From the **Slab Material** drop-down, select the desired material for slab.
- Click on the **Modify/Show Notional Size** button from the dialog box. The **Time Dependent Parameters** dialog box will be displayed; refer to Figure-33. Specify the desired value of notional size in the edit box after selecting the **User-defined** radio button. You can select the **Auto** radio button if you want this value to be specified automatically. In the **Factor** edit box, specify the desired multiplication factor for notional size. Click on the **OK** button from the dialog box to apply the parameters.

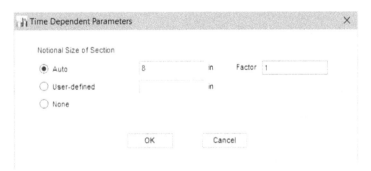

Figure-33. Time Dependent Parameters dialog box

- Select the desired option in the **Modeling Type** drop-down to define how slab will be modeled in ETABS. Select the **Shell-Thin** option if you want to neglect transverse shear deformation and make your model follow Kirchhoff application. Select the **Shell-Thick** option to create a model which follows Mindlin/Reissner application and considers transverse shear deformation. Select the **Membrane** option if you want to model a slab which transfers all the load to supporting structure without taking any deformation. Select the **Layered** option if you want to model a slab with different materials layered in it. This option enables to create mixed material composite slabs. After selecting the **Layered** option, click on the **Modify/Show Layered Slab Data** button. The **Slab Property Layer Definition Data** dialog box will be displayed; refer to Figure-34. Click on the **Add** button in the dialog box. A layer will be added for the current slab. Modify the parameters in the table and add the other layers as required by using the **Add** button. After setting the other parameters, click on the **OK** button. Note that if you have selected **Shell-Thin**, **Shell-Thick**, or **Membrane** option in the **Modeling Type** drop-down then you will be asked to specify the thickness and type of slab in the **Property Data** area of the dialog box; refer to Figure-35.

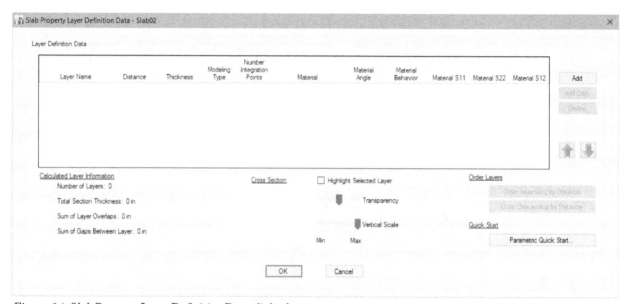

Figure-34. Slab Property Layer Definition Data dialog box

Figure-35. Property Data area

• Select the desired slab type option from the **Type** drop-down and specify the related size parameters.

- Similarly, set the other parameters in the **Slab Property Data** dialog box and click on the **OK** button. The **Slab Properties** dialog box will be displayed again.

You can use the other tools in the **Slab Properties** dialog box in the same way as discussed earlier. Click on the **OK** button from the dialog box exit.

DEFINING DECK SECTIONS

Deck is a flat surface similar to floor but constructed outdoor generally elevated from ground. The procedure to define section properties of deck is given next.

- Click on the **Deck Sections** tool from the **Section Property** cascading menu in the **Define** menu. The **Deck Properties** dialog box will be displayed; refer to Figure-36.

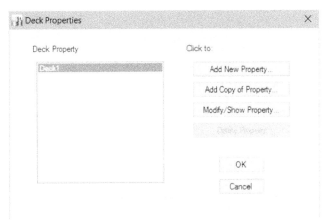

Figure-36. Deck Properties dialog box

- Click on the **Add New Property** tool from the dialog box. The **Deck Property Data** dialog box will be displayed; refer to Figure-37.
- Set the desired parameters for the deck section and click on the **OK** button from the dialog box.

Figure-37. Deck Property Data dialog box

- You can use the other options in the **Deck Properties** dialog box in the same way as discussed earlier.

DEFINING WALL SECTION

The **Wall Sections** tool in the **Section Properties** cascading menu of the **Define** menu is used to define section properties for wall. The procedure to use this tool is same as discussed for the other tools in the **Section Properties** cascading menu.

DEFINING REINFORCING BAR SIZES

The **Reinforcing Bar Sizes** tool is used to define or modify the bar sizes used in reinforcing concrete. The procedure to use this tool is given next.

- Click on the **Reinforcing Bar Sizes** tool from the **Section Properties** cascading menu of the **Define** menu. The **Reinforcing Bar Sizes** dialog box will be displayed; refer to Figure-38.

Figure-38. Reinforcing Bar Sizes dialog box

- Click on the **Add Common Bar Set** button to add the standard bar sizes in the list. The **Select Common Rebar Set** dialog box will be displayed; refer to Figure-39.

Figure-39. Select Common Rebar Set
dialog box

- Select the desired option from the dialog box and click on the **OK** button to add the selected rebar set in the list.
- To manually add a rebar id and size, click twice with a pause in the empty field of the table and specify the desired parameters.
- To modify any field, click twice with a pause on it and set the desired value.
- Click on the **Clear All Bars** button from the dialog box to clear the table.
- Click on the **OK** button from the dialog box to apply the changes.

DEFINING LINK/SUPPORT PROPERTIES

Link is used to join two points separated by some distance. The **Link/Support Properties** tool is used to define link or support properties. The procedure to use this tool is given next.

- Click on the **Link/Support Properties** tool in the **Section Properties** cascading menu of the **Define** menu. The **Define Link Properties** dialog box will be displayed; refer to Figure-40.

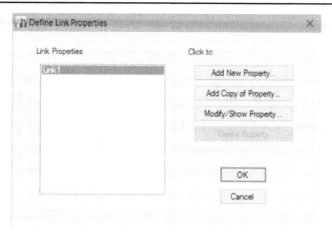

Figure-40. Define Link Properties dialog box

- Click on the **Add New Property** button from the dialog box. The **Link Property Data** dialog box will be displayed; refer to Figure-41.

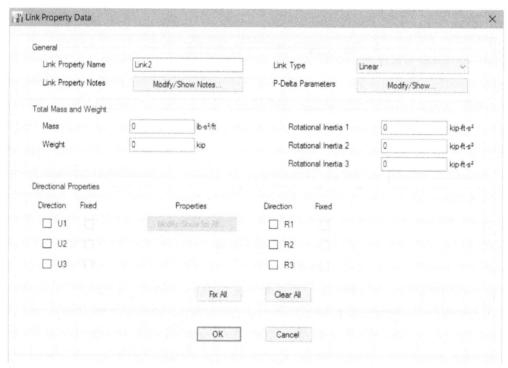

Figure-41. Link Property Data dialog box

- Select the desired option from the **Link Type** drop-down. There are various options in this drop-down to define links for analysis purposes like Linear, Damper-Exponential, Damper-Bilinear and so on.
- Set the other properties of link as required like mass, weight, inertia, and directional properties.
- Click on the **Modify/Show** button for **P-Delta Parameters** in the dialog box to define parameters for P-delta combinations of load. The **Advanced Link P-Delta Parameters** dialog box will be displayed; refer to Figure-42.

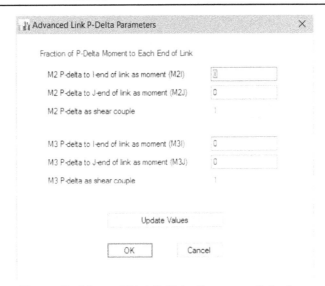

Figure-42. Advanced Link P-Delta Parameters dialog box

- Set the desired parameters in the dialog box to define P-Delta load equation. Note that you should use the P-Delta iteration only when you are concerned about local buckling of structural members. Otherwise you should avoid it because it takes a lot of computing power to solve these iterations.
- Click on the **OK** button from the **Link Property Data** dialog box to create the link property. The **Define Link Properties** dialog box will be displayed again with new link property in the list.

Similarly, you can use the other tools in the **Define Link Properties** dialog box. Click on the **OK** button after performing required changes.

DEFINING FRAME/WALL HINGE PROPERTIES

The Frame/wall hinges are used to provide joints in different frame/ wall sections. The procedure to define these properties is given next.

- Click on the **Frame/Wall Nonlinear Hinges** tool from the **Section Properties** cascading menu of the **Define** menu. The **Define Frame/Wall Hinge Properties** dialog box will be displayed; refer to Figure-43.
- Click on the **Add New Property** button from the dialog box to create a new property. The **Default For Added Hinges** dialog box will be displayed; refer to Figure-44.

Figure-43. Define Frame Wall Hinge Properties dialog box

Figure-44. Default For Added Hinges dialog box

- Select the desired radio button from the dialog box and click on the **OK** button. The **Hinge Property Data** dialog box will be displayed; refer to Figure-45.

Figure-45. Hinge Property Data dialog box

- Specify the property name in the **Hinge Property Name** edit box.
- Set the desired radio button from the **Hinge Type** area of the dialog box to define whether the hinge is brittle or ductile.

- Click in the drop-down and select the desired option (like we have selected Torsion T with ductile function). The related dialog box will be displayed; refer to Figure-46.

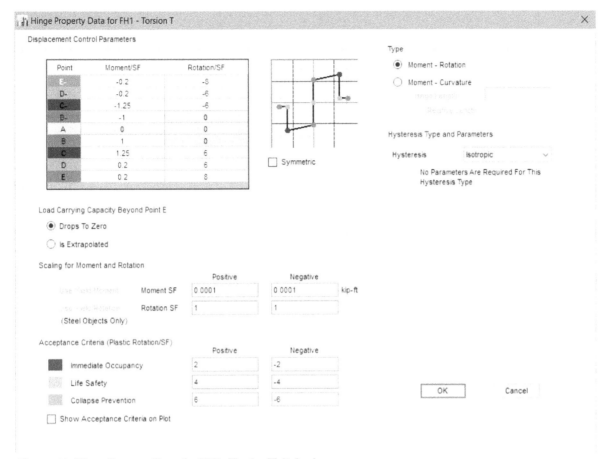

Figure-46. Hinge Property Data for FH1-Torsion T dialog box

- Enter the desired material parameters in the dialog box and click on the **OK** button. The **Hinge Property Data** dialog box will be displayed again.
- Click on the **OK** button from the dialog box to create it. The new property will get added in the list.

DEFINING PANEL ZONE

Panel zone in ETABS is used to define the flexibility between beam-column connection. The procedure to define panel zone properties is given next.

- Click on the **Panel Zone** tool from the **Section Properties** cascading menu in the **Define** menu. The **Panel Zone Properties** dialog box will be displayed; refer to Figure-47.

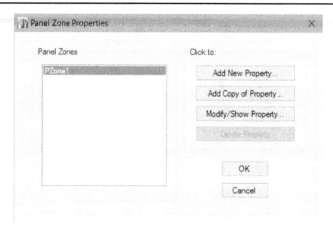

Figure-47. Panel Zone Properties dialog box

- Click on the **Add New Property** button from the dialog box. The **Panel Zone Data** dialog box will be displayed; refer to Figure-48.

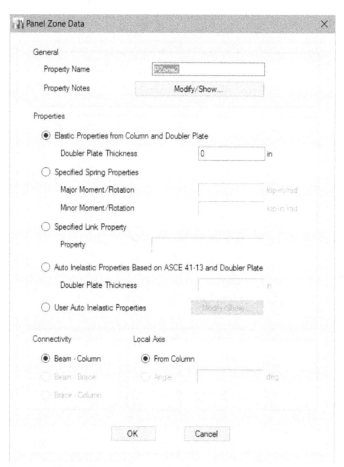

Figure-48. Panel Zone Data dialog box

- Specify the desired name for the property.
- Select the desired radio button from the **Properties** area and specify the related parameters.
- Click on the **OK** button to created the property. The **Panel Zone Properties** dialog box will be displayed again.
- You can use the other buttons in this dialog box as discussed earlier. Click on the **OK** button to exit the dialog box.

DEFINING SPRINGS PROPERTIES

The tools to define properties for different type of springs are available in the **Spring Properties** cascading menu of the **Define** menu; refer to Figure-49. The tools in this cascading menu are discussed next.

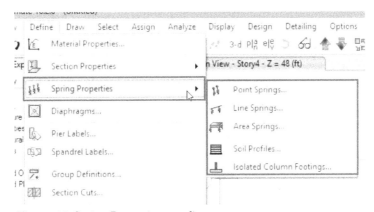

Figure-49. Spring Properties cascading menu

Defining Point Spring Properties

The procedure to define point spring properties is given next.

- Click on the **Point Springs** tool from the **Spring Properties** cascading menu. The **Point Spring Properties** dialog box will be displayed; refer to Figure-50.

Figure-50. Point Spring Properties dialog box

- Click on the **Add New Property** button from the dialog box. The **Point Spring Property Data** dialog box will be displayed; refer to Figure-51.
- Specify the desired name for the spring property and set the other general data.

Figure-51. Point Spring Property Data dialog box

- Set the desired stiffness parameters in different edit boxes of the **Simple Spring Stiffness in Global Directions** area for different directions.
- Click on the **Add** button from the **Single Joint Links at Point** area. A link property will be added.
- Set the desired parameter for link property and click on the **OK** button.
- You can use the other parameters of the **Point Spring Properties** dialog box in the same way as discussed earlier.

Defining Line Spring and Area Spring Properties

The procedures to create line spring and area spring properties are similar to as discussed earlier for Point springs. In case of Line spring properties, you can create properties for an array of spring in straight line. In case of Area spring properties, you can create properties of a full area to be used as spring.

Soil Profile

The soil properties are important factors for construction of a building. The procedure to define soil profile in ETABS is given next.

- Click on the **Soil Profile** tool from the **Spring Properties** cascading menu of the **Define** menu. The **Soil Profiles** dialog box will be displayed; refer to Figure-52.

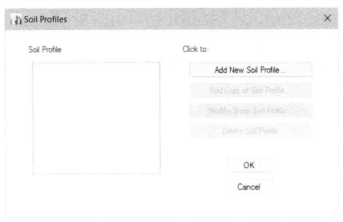

Figure-52. Soil Profiles dialog box

• Click on the **Add New Soil Profile** button from the dialog box. The **Soil Profile Data** dialog box will be displayed; refer to Figure-53.

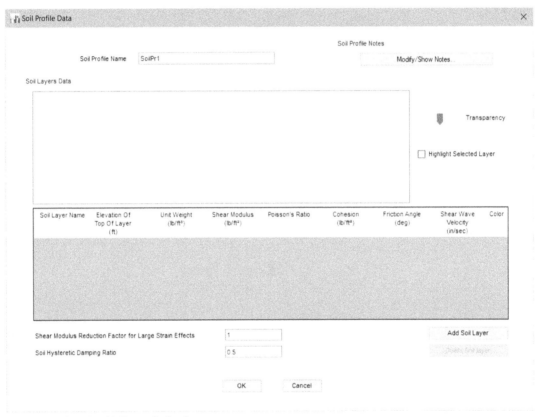

Figure-53. Soil Profile Data dialog box

• Set the desired name for the property in the **Soil Profile Name** edit box.
• Click on the **Add Soil Layer** button. A new layer of soil will be added in the table.
• Triple-click in the field of table to edit it. You can add as many soil layers as needed and edit their parameters.
• After setting the desired properties, click on the **OK** button.

The other tools in the **Soil Profiles** dialog box are same as discussed earlier.

Properties of Isolated Column Footing

The Isolated column footing is used to support single column in the building. The procedure to define properties of isolated column footing is given next.

- Click on the **Isolated Column Footings** tool from the **Spring Properties** cascading menu of the **Define** menu. The **Isolated Column Footing Specifications** dialog box will be displayed; refer to Figure-54.

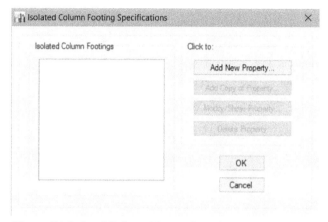

Figure-54. Isolated Column Footing Specifications dialog box

- Click on the **Add New Property** button. The **Isolated Column Footing Data** dialog box will be displayed; refer to Figure-55.

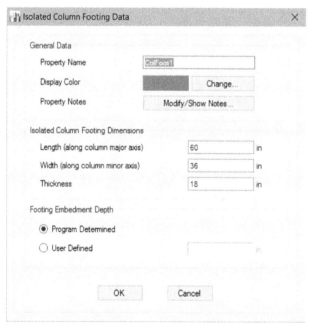

Figure-55. Isolated Column Footing Data dialog box

- Specify the desired parameters and click on the **OK** button to create the property.

You can use the other tools in the **Isolated Column Footing Specifications** dialog box as discussed for other properties.

DIAPHRAGMS PROPERTIES

Diaphragm is used to transfer lateral loads to vertical resisting members of structure. The procedure to create diaphragm property is given next.

• Click on the **Diaphragms** tool from the **Define** menu. The **Define Diaphragm** dialog box will be displayed; refer to Figure-56.

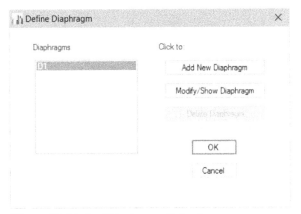

Figure-56. Define Diaphragm dialog box

• Click on the **Add New Diaphragm** button from the dialog box. The **Diaphragm Data** dialog box will be displayed; refer to Figure-57.

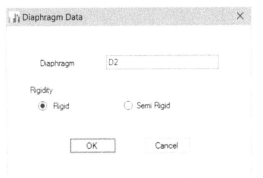

Figure-57. Diaphragm Data dialog box

• Set the desired name and set the desired parameters. Click on the **OK** button.

You can use the other options in the **Define Diaphragm** dialog box in the same way as discussed for other properties dialog boxes.

PIER LABELS AND SPANDREL LABELS

Pier is the load bearing member of structure generally used to support arcs or middle section of bridges. Spandrel is the frame member used to carry load of exterior walls. The procedure to create pier label is given next. You can apply similar procedure for spandrel labels.

• Click on the **Pier Labels** tool from the **Define** menu. The **Pier Labels** dialog box will be displayed; refer to Figure-58.

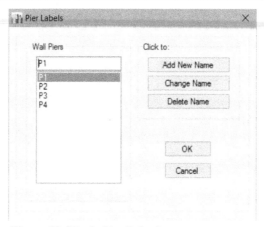

Figure-58. Pier Lables dialog box

- Specify the desired name in the **Wall Piers** edit box and click on the **Add New Name** button. A new pier label will be added.
- If you want to change the name of a pier label, select it from the list and click on the **Change Name** button.
- To delete any label, select it from the list and click on the **Delete Name** button.
- Click on the **OK** button from the dialog box to apply the modifications.

CREATING GROUPS

Groups are used to combine desired objects to form one type of system. The procedure to create group is given next.

- Click on the **Group Definitions** tool from the **Define** menu. The **Define Groups** dialog box will be displayed; refer to Figure-59.

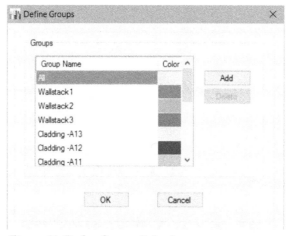

Figure-59. Define Groups dialog box

- Click on the **Add** button to define a new group.
- Triple-click on the name of group to be renamed. Click on the color block next to group name to change it.
- Click on the **OK** button from the **Define Groups** dialog box to create/ modify the groups.

This concludes the properties generally defined regarding the objects in ETABS. But this does not end the tool array of ETABS. In the next chapter, you will learn to assign these properties and many others to the objects in the model.

FOR STUDENT NOTES

Chapter 4

Assigning Properties and Applying Loads

The major topics covered in this chapter are:

- *Introduction*
- *Assigning Properties to Joints*
- *Assigning Frame Properties*
- *Assigning Shell Properties*
- *Assigning Link Properties*
- *Assigning Properties to Tendons*
- *Assigning Joint Loads*
- *Assigning Frame Loads*
- *Assigning Shell Loads*
- *Assigning Loads to Tendons*
- *Copying and Pasting Assigns*

INTRODUCTION

In previous chapter, you have learned to define various properties of frame sections, springs, diaphragms etc. In this chapter, you will learn to assign these properties and many other to the frame members in the model. You will also learn to apply load and prepare model for different analyses.

ASSIGNING PROPERTIES TO JOINTS

Joints are the places where two or more frame members join together. Depending on the joint made at this place, you need to assign the related properties for the analysis. The tools to assign joint properties are available in the **Joint** cascading menu of the **Assign** menu; refer to Figure-1. The tools in this cascading menu are discussed next.

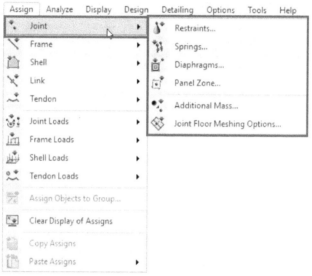

Figure-1. Joint cascading menu

Assigning Restraint Joint Properties

The **Restraint** tool in **Joint** cascading menu is used to restrict the motion of selected joint in specified directions. The procedure to use this tool is given next.

- Select the desired joints from the model to which you want to assign restraints; refer to Figure-2 and click on the **Restraints** tool from the **Joint** cascading menu of the **Assign** menu. The **Joint Assignment - Restraints** dialog box will be displayed; refer to Figure-3.
- Select the check boxes for the directions which you want to be restrained from the **Restraints in Global Directions** area of the dialog box.
- There are four buttons in the **Fast Restraints** area to quickly apply restraints viz. Fixed Base button ⊥, Pinned Base button ▲, Roller Support button ▲, and No Support button •. Select the Fixed Base button if you want to restrict motion in all six directions for selected joints. Select the Pinned Base button if you want to restrict the translation of joints but they are free to rotate.

Select the Roller Support button if you want to restrict the movement of joint along Z axis only. Select the No Support button if you want the joints to be free to move in all directions.

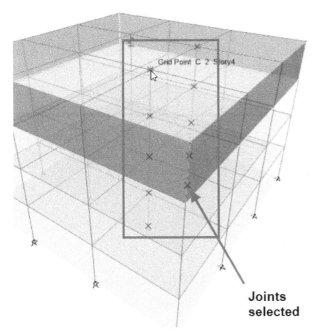

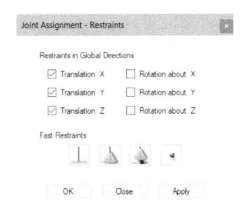

Figure-3. Joint Assignment-Restraints dialog box

Joints selected

Figure-2. Selecting joints

- After setting the desired restraints, click on the **OK** button from the dialog box to apply changes.

Assigning Spring Joint Properties

The **Spring** tool in **Joint** cascading menu is used to assign spring properties to the selected joints. Assigning spring joint is also useful when the beams/columns are joined by soft materials to absorb shocks. The procedure to use this tool is given next.

- Select the joints to which you want to assign spring properties and click on the **Spring** tool from the **Joint** cascading menu of the **Assign** menu. The **Joint Assignment - Spring** dialog box will be displayed with list of spring properties earlier created; refer to Figure-4. The procedure to create spring properties has been discussed in previous chapter.

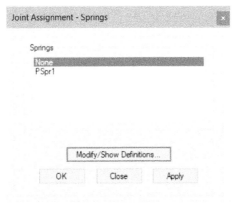

Figure-4. Joint Assignment-Springs dialog box

- Select the desired property and click on the **Modify/Show Definitions** button to check or modify the selected properties. The **Point Spring Properties** dialog box will be displayed; refer to Figure-5. The options in this dialog box have been discussed earlier in previous chapter. Create or modify the properties as required and click on the **OK** button.

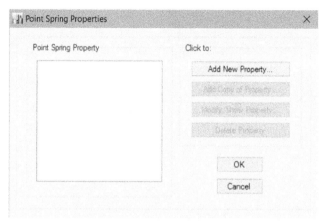

Figure-5. Point Spring Properties dialog box

- Select the desired property from the list in **Joint Assignment - Spring** dialog box and click on the **OK** button to assign properties.

Assigning Diaphragms Joint Properties

A diaphragm joint creates links between joints located within a plane such that they move together as a planar diaphragm, rigid against membrane (in-plane) deformation, but susceptible to plate (out-of-plane) deformation and associated effects. The **Diaphragms** tool in the **Joint** cascading menu is used to assign diaphragms properties to the selected joints. The procedure to use this tool is given next.

- After selecting the joints, click on the **Diaphragms** tool from the **Joint** cascading menu of the **Assign** menu. The **Joint Assignment - Diaphragms** dialog box will be displayed; refer to Figure-6.

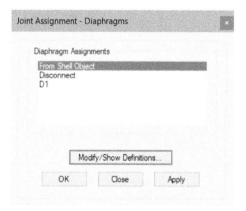

Figure-6. Joint Assignment–Diaphragms dialog box

- List of the all the diaphragm properties earlier created will be displayed in the list box. The procedure to create and modify these properties has already been discussed in previous chapter.

Select the desired property and click on the **OK** button to assign the property to selected joints.

Assigning Panel Zone Properties to Joints

Panel zones are assigned to joints to model the flexibility of beam-column connections. Like you can define the translation and rotation values for the selected connections. The procedure to assign panel zone to selected joints is given next.

• Select the desired joints and click on the **Panel Zone** tool from the **Joint** cascading menu of the **Assign** menu. The **Joint Assignment-Panel Zone Property** dialog box will be displayed; refer to Figure-7.

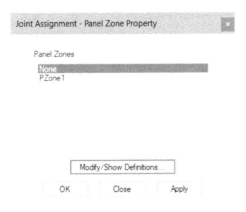

Figure-7. Joint Assignment-Panel Zone Property dialog box

• Select the desired panel zone from the list and click on the **OK** button. The selected panel zone will be assigned.

Assigning Additional Mass to Selected Joints

The **Additional Mass** tool in the **Joint** cascading menu is used to assign additional mass to the selected joints. This additional mass can be representation of any physical object like a machine or furniture etc. The procedure to use this tool is given next.

• Select the joints to which you want to assign additional mass and click on the **Additional Mass** tool from the **Joint** cascading menu of the **Assign** menu. The **Joint Assignment-Additional Mass** dialog box will be displayed; refer to Figure-8.

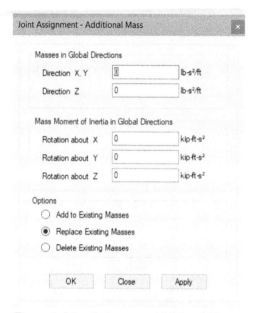

*Figure-8. Joint Assignment–Additional Mass
dialog box*

- Specify the desired values of mass and mass moment of inertia in respective edit boxes.
- Select the desired option from the **Options** area of the dialog box. Select the **Add to Existing Masses** radio button to add the specified value of mass to existing masses. Select the **Replace Existing Masses** radio button to replace the values of masses for selected joints. Select the **Delete Existing Masses** radio button to delete the masses of selected joints for analysis.
- After setting the desired parameters, click on the **OK** button from the dialog box.

Setting Joint Floor Meshing Options

The **Joint Floor Meshing Options** tool is used to set the parameters for meshing of joints. The procedure to use this tool is given next.

- Click on the **Joint Floor Meshing Options** tool from the **Joint** cascading menu of the **Assign** menu. The **Joint Assignment-Joint Floor Meshing Option** dialog box will be displayed; refer to Figure-9.

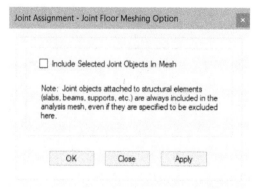

*Figure-9. Joint Assignment-Joint Floor Meshing
Option dialog box*

- Select the **Include Selected Joint Objects In Mesh** check box if you want to include the selected joints in meshing.
- Click on the **OK** button from the dialog box to apply the changes.

ASSIGNING FRAME PROPERTIES

There are various tools in the **Frame** cascading menu of the **Assign** menu to assign different properties to selected frame members; refer to Figure-10. These tools are discussed next.

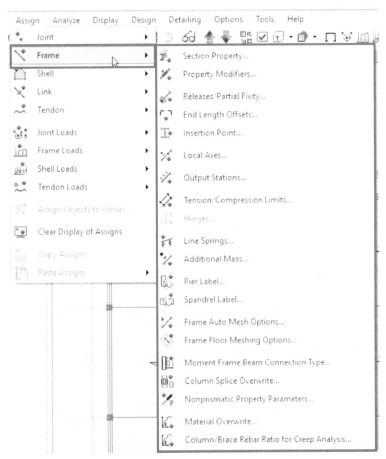

Figure-10. Frame cascading menu

Assigning Section Properties

Section properties are the properties of frame member to define shape, size, and so on. The procedure to assign section property is given next.

- Select the desired frame member and click on the **Section Property** tool from the **Frame** cascading menu of the **Assign** menu. The **Frame Assignment - Section Property** dialog box will be displayed; refer to Figure-11.

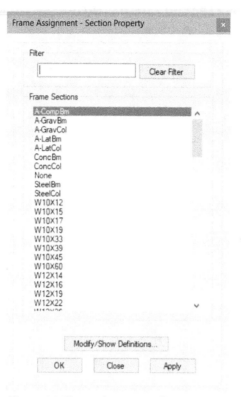

Figure-11. Frame Assignment-Section Property dialog box

- Select the desired option from the list to assign respective property to selected members.
- Click on the **Modify/Show Definitions** tool from the dialog box to check or modify the property of selected section. The procedure to modify or create section properties has been discussed earlier in previous chapter.
- After selecting the desired section property, click on the **OK** button to assign the selected property.

Assigning Property Modifier to Frame

The property modifier is used to modify properties of selected sections exclusively for analysis. The procedure to apply property modifier is given next.

- Select the section whose properties are to be modified for analysis and click on the **Property Modifiers** tool from the **Frame** cascading menu of the **Assign** menu. The **Frame Assignment-Property Modifiers** dialog box will be displayed; refer to Figure-12.

Figure-12. Frame Assignment-Property Modifiers
dialog box

• Specify the desired values in the edit boxes of the dialog box and click on the **OK** button. The specified values will be assigned to selected frame members.

Releasing or Partially Fixing Ends of Frame Members

The **Releases/Partially Fixity** tool is used to release or partially fix frame members for analysis. The procedure to use this tool is given next.

• Select the frame members to which you want to apply releasing or partially fixing and click on the **Releases/Partially Fixity** tool from the **Frame** cascading menu of the **Assign** menu. The **Frame Assignment-Releases/Partial Fixity** dialog box will be displayed; refer to Figure-13.

Figure-13. Frame Assignment-Releases Partial Fixity dialog box

• Select the **Start** or **End** check boxes for the desired type of loads to be released. On selecting the **Start** check box, you release the respective load from start of the frame member and on selecting the **End** check box, you release the load from the end of the frame member.

- If you want to apply partial load on the selected end of frame then specify the desired value in the edit box next to selected check box.
- After setting the desired parameters, click on the **OK** button from the dialog box.

Assigning End Length Offset

The **End Length Offset** tool is used to offset extra length of beam/column at an end point on the joint so that the beam/columns extends at the joint like it happens in real world. Otherwise, the software will join the beams/columns at the centroid. The procedure to use this tool is given next.

- Select the beams/columns to which you want to assign offset and click on the **End Length Offset** tool from the **Frame** cascading menu of the **Assign** menu. The **Frame Assignment-End Length Offsets** dialog box will be displayed; refer to Figure-14.

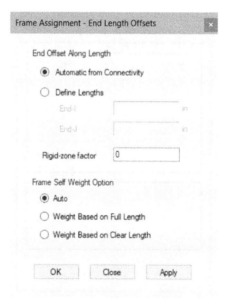

*Figure-14. Frame Assignment-End
Length Offsets dialog box*

- Select the **Automatic from Connectivity** radio button to automatically offset one beam/column to accommodate other beam/column in the joint.
- Select the **Define Lengths** radio button to specify the desired values of offset along End-I and End-J of the beam/column. End-I is the start point of beam/column and End-J is the end point of the beam/column.
- Specify the desired value in **Rigid-zone factor** edit box. The rigid-zone factor specifies the fraction of each end offset assumed to be rigid for bending and shear deformations. When a fraction of the end offset is specified rigid, the outside portion of the end offset is assumed rigid, that is, the portion at the end of the frame member.
- Select the desired option from the **Frame Self Weight Option** area to

define how the weight of the frame member will be calculated after assigning the offset.
• After setting the desired parameters, click on the **OK** button.

Assigning Insertion Points for Frame Members

The insertion point is the location of beam/column face at which two more beams/columns join together. The **Insertion Point** tool in the **Frame** cascading menu of the **Assign** menu is used to specify insertion points for selected beams/columns. The procedure to use this tool is given next.

• After selecting desired beams/columns, click on the **Insertion Point** tool from the **Frame** cascading menu of the **Assign** menu. The **Frame Assignment-Insertion Point** dialog box will be displayed; refer to Figure-15.

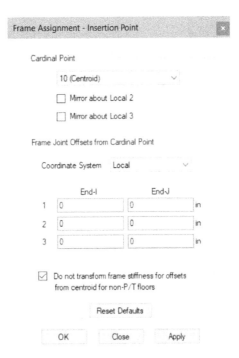

Figure-15. Frame Assignment-Insertion Point dialog box

• Select the desired option from the **Cardinal Point** drop-down to define the location of insertion point. Location of various points on the beam is shown in Figure-16.

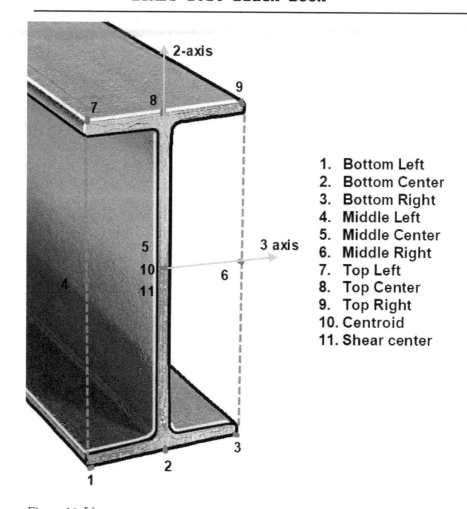

1. **Bottom Left**
2. **Bottom Center**
3. **Bottom Right**
4. **Middle Left**
5. **Middle Center**
6. **Middle Right**
7. **Top Left**
8. **Top Center**
9. **Top Right**
10. **Centroid**
11. **Shear center**

Figure-16. I beam

- Select the **Mirror about 2** and/or **Mirror about 3** check boxes to mirror the cardinal point about 2 axis and/or 3 axis.
- Specify the desired offset values in the edit boxes of **Frame Joint Offsets from Cardinal Point** area of the dialog box.
- After setting the desired parameters, click on the **OK** button from the dialog box.

Assigning Local Axes

The **Local Axes** tool in **Assign** menu is used to set the orientation of axis of selected frame member. The procedure to use this tool is given next.

- Click on the **Local Axes** tool from the **Frame** cascading menu of the **Assign** menu. The **Frame Assignment-Local Axes** dialog box will be displayed; refer to Figure-17.
- Select the **Angle** radio button to specify the orientation of 2nd axis of the selected frame member. Select the **Rotate by Angle** radio button to specify the angle value by which the 2nd axis will be rotated with respect to 1st axis.
- Select the **Orient with Grid System - Applies to Vertical Frame Objects Only** radio button to orient the axes by using grid system. Note that these options are applicable for vertical frame objects only like columns.

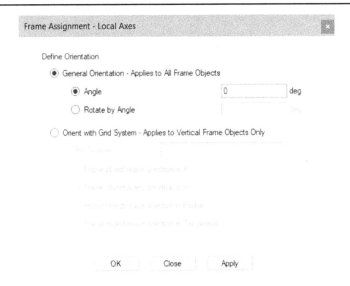

Figure-17. Frame Assignment-Local Axes dialog box

- Select the desired option below the selected radio button to specify the orientation of selected frame members with respect to grid system selected.
- After setting the desired parameters, click on the **OK** button. The marking of axes will be displayed on the model; refer to Figure-18.

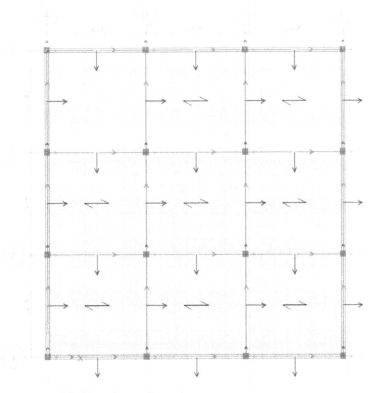

Figure-18. Model with axes denotation

Assigning Output Stations

Output stations are designated locations along a frame element used for output forces, perform design, and plot points used for graphic display of force diagrams. The procedure to assign output stations to selected frame members is given next.

- After selecting the frame members, click on the **Output Stations** tool from **Frame** cascading menu of the **Assign** menu. The **Frame Assignment-Output Stations** dialog box will be displayed; refer to Figure-19.

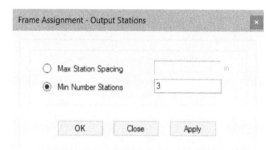

Figure-19. Frame Assignment-Output Stations dialog box

- Select the **Max Station Spacing** radio button to specify the maximum distance between two consecutive stations being placed on selected frame members. Specify the desired distance value in the adjacent edit box.
- Select the **Min Number Stations** radio button and specify the minimum number of stations to be assigned on selected frame members.
- Click on the **OK** button to apply changes.

Assigning Tension/Compression Limits

The tension and compression limits are the maximum allowed tension and compression stresses on the selected frame members. These options are used primarily to model tension-only cables and braces or compression only beams. The procedure to assign tension/compression limits is given next.

- After selecting the frame members, click on the **Tension/Compression Limits** tool from the **Frame** cascading menu of the **Assign** menu. The **Frame Assignment-Tension/Compression Limits** dialog box will be displayed; refer to Figure-20.

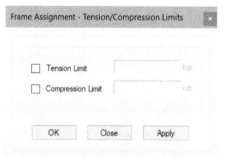

Figure-20. Frame Assignment-Tension Compression Limits dialog box

- Select the **Tension Limit** check box to specify tension limit in selected frame members and specify the value in adjacent edit box. Note that the unit for tension limit is kip which is 1000 pound-force.

- Select the **Compression Limit** check box to specify compression limit in selected frame members and specify the value in adjacent edit box. Note that the unit for compression limit is kip which is 1000 pound-force.
- After setting the desired values, click on the **OK** button to apply limits.

Assigning Hinges

The **Hinges** tool of **Frame** cascading menu in **Assign** menu is used to assign hinge properties to the selected frame member. The procedure to use this tool is given next.

- After selecting the frame member, click on the **Hinges** tool from the **Frame** cascading menu of the **Assign** menu. The **Frame Assignment-Hinges** dialog box will be displayed; refer to Figure-21.

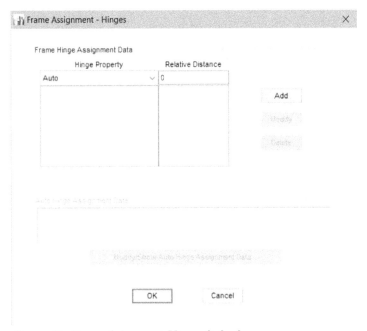

Figure-21. Frame Assignment-Hinges dialog box

- Specify the relative distance of hinge at the selected frame member in **Relative Distance** edit box of the dialog box. The value specified here should be between 0 to 1. Here, 0 and 1 represent the end points of selected frame member and 0.5 represents the middle of frame member. After setting the desired value, click on the **Add** button. The **Auto Hinge Assignment Data** dialog box will be displayed; refer to Figure-22.
- Select the desired option from **Auto Hinge Type** drop-down to define the source of list from where hinges will be selected. If you have selected the **Buckling Restrained Brace** option from the drop-down then hinges will be selected automatically based on braces created in the model.
- Select the desired table from the **Select a Hinge Table** drop-down. Note that the selection of table depends on type and material of selected frame member.

- Specify the degree of freedom and loading parameters as required and then click on the **OK** button to apply automatic hinge. Based on material and loading requirements, the hinge will be created automatically.

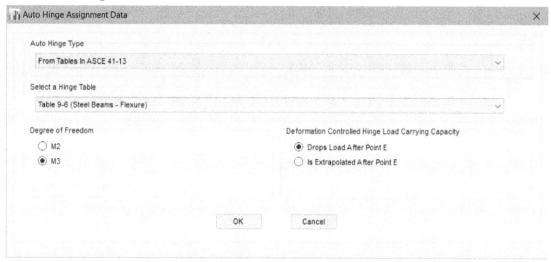

Figure-22. Auto Hinge Assignment Data dialog box

Assigning Hinge Overwrites

The **Hinge Overwrites** tool is used to assign overwrite to the selected frame member. These overwrites include the sub-division of hinge and no load drop on hinges. The procedure to use this tool is given next.

- After selecting the frame members, click on the **Hinge Overwrites** tool from the **Frame** cascading menu of the **Assign** menu. The **Frame Assignment-Hinge Overwrites** dialog box will be displayed; refer to Figure-23.

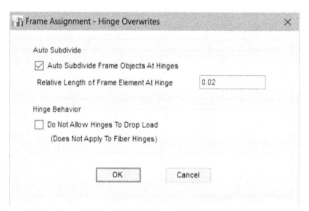

Figure-23. Frame Assignment-Hinge Ovewrites dialog box

- Select the **Auto Subdivide Frame Objects At Hinges** check box to divide the frame objects on which hinge point exist. You can specify the value of relative length in the edit box as discussed for previous tool.
- Select the **Do Not Allow Hinges To Drop Load** check box to make sure the hinged frame members deform as per their stiffness and do not fail suddenly.

- After setting the desired parameters, click on the **OK** button.

Assigning Line Spring

The **Line Spring** tool in the **Frame** cascading menu of the **Assign** menu is used to assign line spring properties to selected frame members. The procedure to use this tool is given next.

- Select the desired frame members and click on the **Line Spring** tool from the **Frame** cascading menu of the **Assign** menu. The **Frame Assignment-Line Springs** dialog box will be displayed; refer to Figure-24.

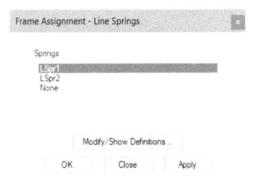

Figure-24. Frame Assignment-Line Springs dialog box

- Select the desired property to be applied.
- Click on the **OK** button to apply the settings.

Assigning Additional Mass to Frame Members

The **Additional Mass** tool in the **Frame** cascading menu of the **Assign** menu is used to assign additional mass to the selected frame members. The procedure to use this tool is given next.

- After selecting the frame members, click on the **Additional Mass** tool from the **Frame** cascading menu of the **Assign** menu. The **Frame Assignment-Additional Mass** dialog box will be displayed; refer to Figure-25.

Figure-25. Frame Assignment-Additional Mass dialog box

- Specify the desired value of mass per length in the **Frame Mass/ Length** edit box.
- Select the desired radio button from the **Options** area to specify whether the mass should be added, replaced, or deleted.
- After setting the desired parameters, click on the **OK** button from the dialog box.

Assigning Pier Labels

The **Pier Label** tool in **Frame** cascading menu of the **Assign** menu is used to assign earlier created pier label to selected frame member. The procedure to use **Pier Label** tool is given next.

- After selecting the desired frame members, click on the **Pier Label** tool from the **Frame** cascading menu of the **Assign** menu. The **Frame Assignment-Pier Label** dialog box will be displayed; refer to Figure-26.

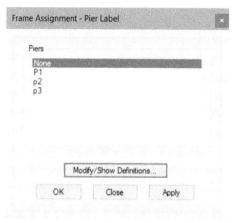

Figure-26. Frame Assignment-Pier Label dialog box

- Select the desired pier label for selected frame member and click on the **OK** button.

You can assign spandrel labels in the same way.

Frame Auto Mesh Options

The **Frame Auto Mesh Options** tool in the **Frame** cascading menu of the **Assign** menu is used to specify whether the auto meshing will be applied to selected frames or not during the analysis. The procedure to use this tool is given next.

- Select the desired frame members and click on the **Frame Auto Mesh Options** tool from the **Frame** cascading menu of the **Assign** menu. The **Frame Assignment-Frame Auto Mesh Options** dialog box will be displayed; refer to Figure-27.

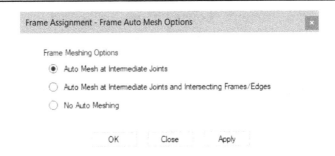

Figure-27. Frame Assignment-Frame Auto Mesh Options dialog box

- Select the desired option from the dialog box. Select the **Auto Mesh at Intermediate Joints** radio button to assign auto meshing at the joints. Select the **Auto Mesh at Intermediate Joints and Intersecting Frames/Edges** radio button to assign auto meshing at joins as well as intersecting frame members and edges. Select the **No Auto Meshing** radio button to exclude selected frame members from auto meshing.
- After setting the desired parameters, click on the **OK** button.

Similarly, you can set the frame floor meshing options by using the **Frame Floor Meshing Options** tool in the **Frame** cascading menu of the **Assign** menu.

Defining Moment Connections of Frame Beams

The **Moment Frame Beam Connection Type** tool in the **Frame** cascading menu of the **Assign** menu is used to define the connection type of the beam related to moment. The procedure to use this tool is given next.

- After selecting the desired beams, click on the **Moment Frame Beam Connection Type** tool from the **Frame** cascading menu of the **Assign** menu. The **Frame Assignment-Moment Frame Beam Connection Type** dialog box will be displayed; refer to Figure-28.

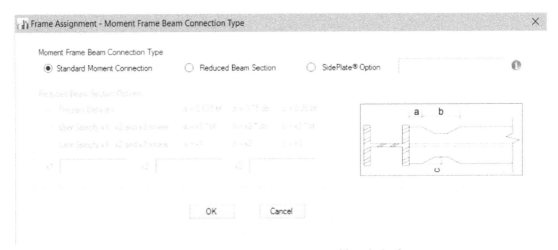

Figure-28. Frame Assignment-Moment Frame Beam Connection Type dialog box

- Select the **Standard Moment Connection** radio button if you want to use standard connections without any section reduction or seismic loading conditions.
- Select the **Reduced Beam Section** radio button if you want reduction in size of the beam near connection point. The options to specify parameters for reduction in beam section will be displayed in the **Reduced Beam Section Options** area of the dialog box. Specify the desired values of a, b, and c using these options.
- Select the **SidePlate Option** radio button if you want to update the connection as per the seismic loading of the joint. In this type of connection, a series of plates are welded to the beam end and the column to strengthen the beam-column connection and force hinging to occur in the beam away from the connection. On selecting this radio button, a drop-down is displayed at the right in the dialog box. Select the desired option from the drop-down to define seismic loading.
- After setting the desired parameters, click on the **OK** button.

Assigning Column Splice Overwrites

The **Column Splice Overwrite** tool is used to apply overwrites to splicing of columns. The procedure to do so is given next.

- Select the desired columns and click on the **Column Splice Overwrite** tool from the **Frame** cascading menu of the **Assign** menu. The **Assign Column Splice Overwrites to Column Objects** dialog box will be displayed; refer to Figure-29.

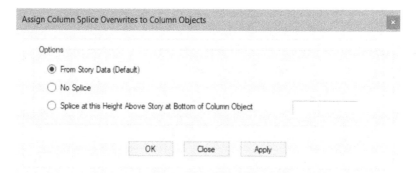

Figure-29. Assign Column Splice Overwrites to Column Objects dialog box

- Select the **From Story Data (Default)** radio button to splice columns based on the story data.
- Select the **No Splice** radio button if you do not want to splice columns.
- Select the **Splice at this Height Above Story at Bottom of Column Object** radio button to specify height above story bottom at which columns will be spliced.

Assigning Nonprismatic Property Parameters

The Nonprismatic Property is the property of frame members by which various parameters of the model like stiffness, weight, axial stress etc. change along the length. The procedure to assign nonprismatic properties is given next.

- Select the desired frame members and click on the **Nonprismatic Property Parameters** tool from the **Frame** cascading menu of the **Assign** menu. The **Frame Assignment - Nonprismatic Property Parameters** dialog box will be displayed; refer to Figure-30.

Figure-30. Frame Assignment-Nonprismatic Property Parameters dialog box

- Select the **Default Parameters** radio button to keep the frame members as they are by default.
- Select the **Advanced User Specified Parameters** radio button to specify the relative length of frame members till which the frame members will not demonstrate nonprismatic properties and enter the desired values in adjacent edit boxes.
- After setting the desired parameters, click on the **OK** button.

Assigning Material Overwrites

The material overwrites are assigned to modify the material properties of selected frame members. The procedure to do so is given next.

- After selecting desired frame members, click on the **Material Overwrites** tool from the **Frame** cascading menu of the **Assign** menu. The **Frame Assignment-Material Overwrites** dialog box will be displayed; refer to Figure-31.

Figure-31. Frame Assignment-Material Overwrite dialog box

- Select the desired material to be assigned to selected frame members and click on the **OK** button.

Assigning Column/Brace Rebar Ratio for Creep Analysis

The rebar ratio is the ratio of steel bars to concrete mixture per volume. The procedure to assign this ratio to selected columns/beams is given next.

- Click on the **Column/Brace Rebar Ratio for Creep Analysis** tool from the **Frame** cascading menu of the **Assign** menu. The **Frame Assignment-Rebar Ratio for Creep Analysis** dialog box will be displayed; refer to Figure-32.

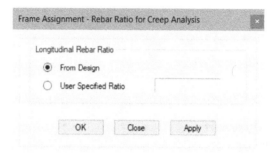

Figure-32. Frame Assignment-Rebar Ratio for Creep Analysis dialog box

- Select the **From Design** radio button if you want to apply rebar ratios as per the design codes automatically.
- Select the **User Specified Ratio** radio button to specify the rebar ratio manually in the adjacent edit box.
- After setting the desired parameters, click on the **OK** button.

ASSIGNING SHELL PROPERTIES

Shell objects are those made of three or four nodes and are used to create floor, wall, bridge deck etc. in the model. The tools to assign various properties to shell objects are available in the **Shell** cascading menu of the **Assign** menu; refer to Figure-33. Various tools in this menu are discussed next.

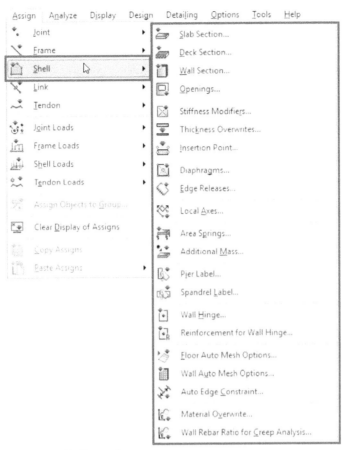

Figure-33. Shell cascading menu

Assigning Slab Section

The **Slab Section** tool in **Shell** cascading menu of the **Assign** menu is used to assign slab section properties to selected shell objects. The procedure to use this tool is given next.

- Select the desired shell objects and click on the **Slab Section** tool from the **Shell** cascading menu of the **Assign** menu. The **Shell Assignment-Slab Section** dialog box will be displayed; refer to Figure-34.

Figure-34. Shell Assignment-Slab Section
dialog box

- Select the desired slab section property from the list and click on the **OK** button from the dialog box. Note that the procedure to create and modify slab section properties has been discussed in previous chapter.

Assigning Deck Section Properties

The **Deck Section** tool in the **Shell** cascading menu of the **Assign** menu is used to assign deck properties to selected shell objects. The procedure to use this tool is given next.

- After selecting the shell object, click on the **Deck Section** tool in the **Shell** cascading menu of the **Assign** menu. The **Shell Assignment-Deck Section** dialog box will be displayed; refer to Figure-35.

Figure-35. Shell Assignment-Deck Section dialog box

- Select the desired deck property and click on the **OK** button. The selected deck property will be assigned shell objects. Note that the procedure to create and modify deck section properties has been discussed in previous chapter.

Assigning Wall Section

The **Wall Section** tool in the **Shell** cascading menu of the **Assign** menu is used assign wall sections to selected shell objects. The procedure to use this tool is given next.

- After selecting the shell objects, click on the **Wall Section** tool in the **Shell** cascading menu of the **Assign** menu. The **Shell Assignment-Wall Section** dialog box will be displayed; refer to Figure-36.
- Select the desired wall section from the dialog box and click on the **OK** button. Note that the procedure to create and modify wall section properties has been discussed in previous chapter.

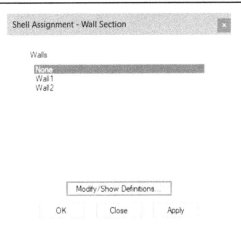

*Figure-36. Shell Assignment-Wall Section
dialog box*

Assigning Openings

The **Opening** tool in the **Shell** cascading menu of the **Assign** menu is used to designate selected shell objects as opening. Openings are assumed to carry no load. The procedure to use this tool is given next.

- After selecting the shell object, click on the **Opening** tool from the **Shell** cascading menu of the **Assign** menu. The **Shell Assignment-Openings** dialog box will be displayed; refer to Figure-37.

*Figure-37. Shell Assignment-Openings
dialog box*

- Select the **Opening** option from the drop-down and click on the **OK** button. The selected shell object will be converted to an opening. If you want to close an opening then select it and use the **Not an Opening** option from the drop-down.

Assigning Stiffness Modifiers

The **Stiffness Modifiers** tool is used to apply modifiers to stiffness properties of the selected shell object. The procedure to use this tool is given next.

- After selecting the desired shell objects, click on the **Stiffness Modifiers** tool from the **Shell** cascading menu of the **Assign** menu. The **Shell Assignment-Stiffness Modifiers** dialog box will be displayed; refer to Figure-38.

Figure-38. Shell Assignment-Stiffness Modifiers dialog box

- Specify the desired values for modifier and click on the **OK** button to apply values.

Assigning Thickness Overwrites

The **Thickness Overwrites** tool is used to modify the thickness of selected shell objects. The procedure to use this tool is given next.

- Select the desired shell objects and click on the **Thickness Overwrites** tool from the **Shell** cascading menu of the **Assign** menu. The **Shell Assignment - Thickness Overwrites** dialog box will be displayed; refer to Figure-39.

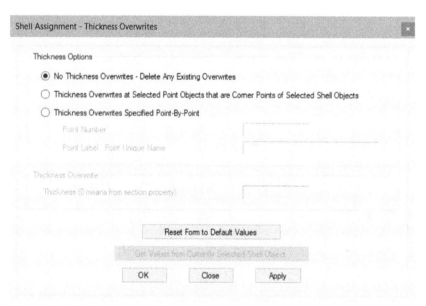

Figure-39. Shell Assignment-Thickness Overwrites dialog box

- Select the **No Thickness Overwrites - Delete Any Existing Overwrites** radio button if you want to delete overwrites of selected shell objects.

- Select the **Thickness Overwrites at Selected Point Objects that are Corner Points of Selected Shell Objects** radio button to overwrite thickness of shell objects at selected points. This option is available only when you have selected shell object and points on it. Specify the value of the thickness in **Thickness** edit box of **Thickness Overwrite** area of the dialog box.
- Select the **Thickness Overwrites Specified Point-By-Point** radio button to specify thickness overwrites for select shell objects point by point. Use the buttons adjacent to selected radio button switch between various buttons. Specify the thickness overwrite values for each point in the **Thickness** edit box.
- After setting desired parameters, click on the **OK** button.

Assigning Insertion Point to Shell Objects

The **Insertion Point** tool in the **Shell** cascading menu of the **Assign** menu is used to specify offset for insertion point of selected shell objects. The procedure to use this tool is given next.

- After selecting desired shell objects, click on the **Insertion Point** tool from the **Shell** cascading menu of the **Assign** menu. The **Shell Assignment-Insertion Point** dialog box will be displayed; refer to Figure-40.

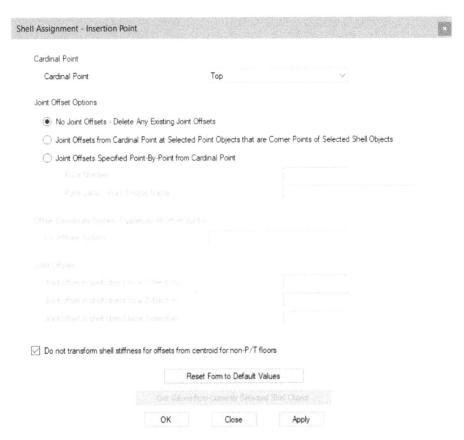

Figure-40. Shell Assignment-Insertion Point dialog box

- Select the location of cardinal point to be used for offsetting from each selected shell object in the **Cardinal Point** drop-down.

- Select the **No Joint Offsets-Delete Any Existing Joint Offsets** radio button to remove any offset applied to selected points of the selected shell objects.
- Select the **Joint Offsets from Cardinal Point at Selected Point Objects that are Corner Points of Selected Shell Objects** radio button to specify the value of offset in each direction using the edit boxes of **Joint Offsets** area of the dialog box. Note that by default, **Local** option is selected in the **Coordinate System** drop-down of the dialog box. If you want to specify the offset along the global X, Y, and Z axes then select the **Global** option from the **Coordinate System** drop-down.
- Select the **Joint Offsets Specified Point-By-Point from Cardinal Point** radio button to specify different offset values for each selected point. To switch between the points, click on the navigation buttons in the dialog box; refer to Figure-41.

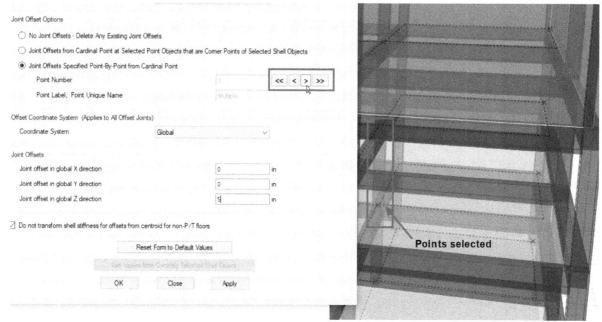

Figure-41. Navigation buttons

- After setting the desired parameters, click on the **OK** button.

Assigning Diaphragms to Shell Objects

The **Diaphragms** tool in the **Shell** cascading menu of the **Assign** menu is used to assign diaphragm property to selected shell objects. The procedure to use this tool is given next.

- Select the desired shell objects and click on the **Diaphragms** tool from the **Shell** cascading menu of the **Assign** menu. The **Shell Assignment - Diaphragms** dialog box will be displayed; refer to Figure-42.

*Figure-42. Shell Assignment-Diaphragms
dialog box*

- Select the desired property from the dialog box and click on the **OK** button.

Assigning Edge Releases

The **Edge Releases** tool is used to release load from edges of selected shell objects. The procedure to use this tool is given next.

- After selecting the desired shell objects, click on the **Edge Releases** tool from the **Shell** cascading menu of the **Assign** menu. The **Shell Assignment - Edge Releases** dialog box will be displayed; refer to Figure-43.

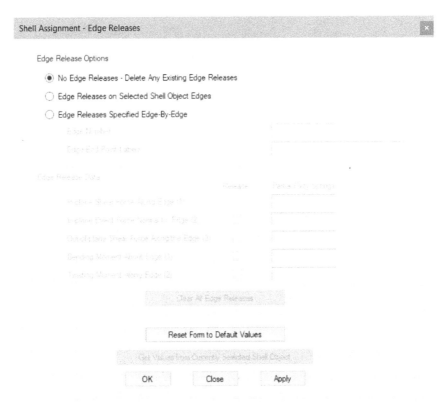

Figure-43. Shell Assignment-Edge Releases dialog box

- Select the **No Edge Releases - Delete Any Existing Edge Releases** radio button to delete any release earlier assigned to selected shell objects.
- Select the **Edge Releases on Selected Shell Object Edges** radio button to apply releases to selected objects. The options in the **Edge Release Data** area of the dialog box will become active. Select the check boxes for loads which you want to be removed. If you want to apply partial load then specify the value in respective edit box of selected check box.
- Select the **Edge Releases Specified Edge-By-Edge** radio button to select the individually set releases for each selected edge. The options are same as discussed earlier.
- After setting the desired parameters, click on the **OK** button from the dialog box.

Assigning Local Axes

The **Local Axes** tool in the **Shell** cascading menu of the **Assign** menu. The procedure to use this tool is given next.

- Select the desired shell objects and click on the **Local Axes** tool from the **Shell** cascading menu of the **Assign** menu. The **Shell Assignment - Local Axes** dialog box will be displayed; refer to Figure-44.

Figure-44. Shell Assignment-Local Axes dialog box

- Select the **Rotation Angle from Default Orientation** radio button if you want to specify angle with respect to default axes and specify the angle value in adjacent edit box.
- Select the **Rotation Angle from Current Orientation** radio button to rotate the axes with respect to current orientation and specify the angle value in adjacent edit box.
- Select the **Normal to Selected Beam** radio button to rotate the local 1 axis perpendicular to selected beam.
- After setting the desired parameters, click on the **OK** button.

Assigning Area Springs

The **Area Spring** tool is used to assign area springs to selected shell objects. The procedure to use this tool is given next.

- Select the desired shell objects and click on the **Area Spring** tool from the **Shell** cascading menu of the **Assign** menu. The **Shell Assignment - Area Spring** dialog box will be displayed; refer to Figure-45.

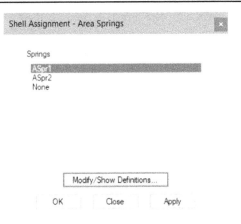

*Figure-45. Shell Assignment-Area Springs
dialog box*

- Select the desired area spring property from the dialog box and click on the **OK** button. The procedure to create and modify area spring properties has been discussed in previous chapter.

Assigning Additional Mass to Shell Objects

The **Additional Mass** tool is used to assign additional mass to selected shell objects. The procedure to assign additional mass is given next.

- Select the shell objects and click on the **Additional Mass** tool from the **Shell** cascading menu of the **Assign** menu. The **Shell Assignment - Additional Mass** dialog box will be displayed; refer to Figure-46.

*Figure-46. Shell Assignment-Addition-
al Mass dialog box*

- Specify the value of mass per area value in the **Shell Mass/Area** edit box.
- Select the desired option from the **Options** area of the dialog box and click on the **OK** button.

The other tools in the **Shell** cascading menu are same as discussed earlier.

ASSIGNING LINK PROPERTIES

The links are used to join two or more joints. The tools in **Link** cascading menu are used to assign link properties to selected links and manage local axes; refer to Figure-47. The tools in this menu are discussed next.

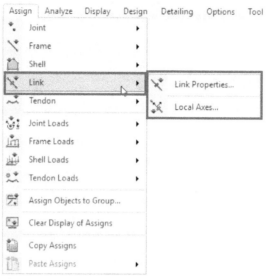

Figure-47. Link cascading menu

Assigning Link Properties

The **Link Properties** tool in the **Link** cascading menu of the **Assign** menu is used to assign link properties earlier created to the selected links. The procedure to use this tool is given next.

• After selecting desired links, click on the **Link Properties** tool from the **Link** cascading menu of the **Assign** menu. The **Link Assignment-Link Property** dialog box will be displayed; refer to Figure-48.

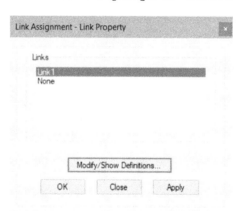

Figure-48. Link Assignment-Link Property dialog box

• Select the desired option from the dialog box and click on the **OK** button. The procedure to create and modify link properties has been discussed earlier.

Orienting Local Axes for the Links

The **Local Axes** tool in the **Link** cascading menu of the **Assign** menu is used to orient axes of selected links. The procedure to use this tool is given next.

- Click on the **Local Axes** tool from the **Link** cascading menu of the **Assign** menu after selecting the links. The **Link Assignment-Local Axes** dialog box will be displayed; refer to Figure-49.

Figure-49. Link Assignment-Local Axes dialog box

- Specify the desired angle value to rotate 1st axis of the selected links.
- Click on the **OK** button from the dialog box to apply specified rotation.

ASSIGNING PROPERTIES TO TENDONS

The tendons are used to prestress other objects in the structure. The procedure to assign properties to tendons has been discussed earlier.

- Select the tendons earlier created in the model and click on the **Tendon Properties** tool from the **Tendon** cascading menu of the **Assign** menu. The **Tendon Property Assign** dialog box will be displayed; refer to Figure-50.

Figure-50. Tendon Property Assign dialog box

- Select the desired tendon property from the **Property Name** drop-down.
- Specify the number of strands of tendon in the **Number of Strands** edit box.
- Select the **Bonded** radio button to bond tendons as if they are modeled with the structure. Select the **Unbounded** radio button to make tendons unbound to the model and created as a separate feature.
- After setting desired parameters, click on the **OK** button.

ASSIGNING JOINT LOADS

The tools in the **Joint Loads** cascading menu are used to assign different types of loads on joints in the model; refer to Figure-51. The tools in this menu are discussed next.

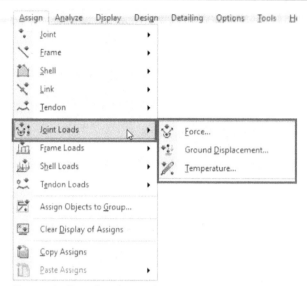

Figure-51. Joint Loads cascading menu

Applying Force at the Joint

The **Force** tool in the **Joint Loads** cascading menu is used to apply force at the joints. The procedure to use this tool is given next.

• Select the joints on which you want to apply loads and click on the **Force** tool from the **Joint Loads** cascading menu of the **Assign** menu. The **Joint Load Assignment-Force** dialog box will be displayed; refer to Figure-52.

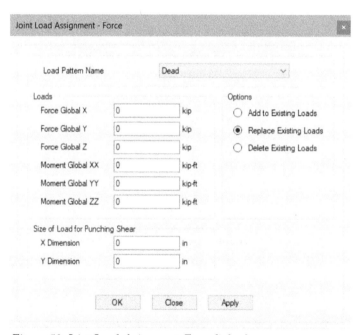

Figure-52. Joint Load Assignment-Force dialog box

• Click in the **Load Pattern Name** drop-down and select the desired type of load. The **Dead** option is selected to apply fixed load (mostly exerted by fixed assets of building) on the joint. Select the **Live** option if you want to apply maximum load of moving objects in the building (like people, moving machines etc.). Select the **PT-FINAL** option to define post tension final load. Select the **PT-TRANSFER** option to define post tension transferred load on selected joints.

- Specify the values of different components of force and moment in respective edit boxes of the **Loads** area.
- Specify the size of punching load in the edit boxes of **Size of Load for Punching Shear** area of the dialog box.
- Select the desired radio button from the **Options** area of the dialog box to add, replace or delete loads.
- After setting the desired parameters, click on the **OK** button from the dialog box.

Applying Ground Displacement at the Joints

The **Ground Displacement** tool is used to apply desired translation or rotation to selected joints. The procedure to use this tool is given next.

- Select the joints to which you want to apply translation or rotation and click on the **Ground Displacement** tool from the **Joint Loads** cascading menu of the **Assign** menu. The **Joint Load Assignment-Ground Displacement** dialog box will be displayed; refer to Figure-53.

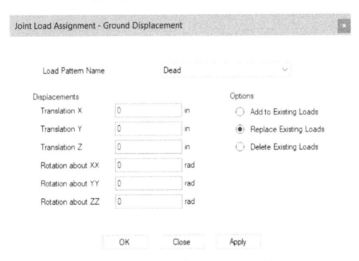

Figure-53. Joint Load Assignment-Ground Displacement dialog box

- Specify the desired values of translation and rotation in respective edit boxes.
- After setting the parameters, click on the **OK** button.

Assigning Temperature to Joints

The **Temperature** tool in the **Joint** cascading menu is used to assign temperature to selected joints. The procedure to use this tool is given next.

- Select the desired joints and click on the **Temperature** tool from the **Joint** cascading menu of the **Assign** menu. The **Joint Load Assignment-Temperature** dialog box will be displayed; refer to Figure-54.

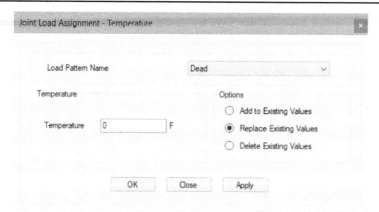

Figure-54. Joint Load Assignment-Temperature dialog box

- Specify the desired temperature value in the **Temperature** edit box. The other options in the dialog box are same as discussed earlier.
- After setting the desired parameters, click on the **OK** button.

ASSIGNING FRAME LOADS

The tools in the **Frame Loads** cascading menu are used to assign loads to frame members; refer to Figure-55. Various tools in this cascading menu are discussed next.

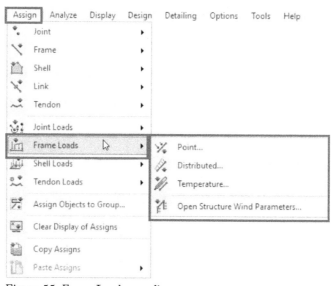

Figure-55. Frame Loads cascading menu

Assigning Point load on Frame Members

The **Point** tool in the **Frame** cascading menu is used to apply point load at frame members. The procedure to use this tool is given next.

- Select the desired frame members and click on the **Point** tool from the **Frame** cascading menu of the **Assign** menu. The **Frame Load Assignment-Point** dialog box will be displayed; refer to Figure-56.

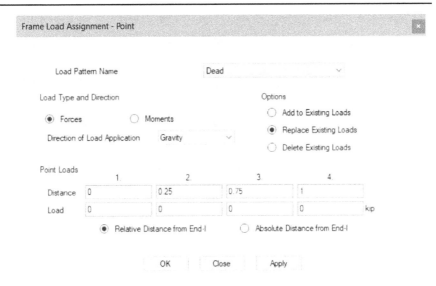

Figure-56. Frame Load Assignment-Point dialog box

- Select the desired radio button to define type of load from the **Load Type and Direction** area of the dialog box.
- Select the desired option from the **Direction of Load Application** drop-down to define direction of load.
- Specify the relative distance of load points and loads on them in the edit boxes of the **Point Loads** area.
- Set the other options as discussed earlier and click on the **OK** button to apply loads.

Assigning Distributed Load on Frame Members

The **Distributed** tool in the **Frame Loads** cascading menu is used to apply uniform load on selected frame members. The procedure to use this tool is given next.

- Select the desired frame members and click on the **Distributed** tool from the **Frame Loads** cascading menu of the **Assign** menu. The **Frame Load Assignment-Distributed** dialog box will be displayed; refer to Figure-57.

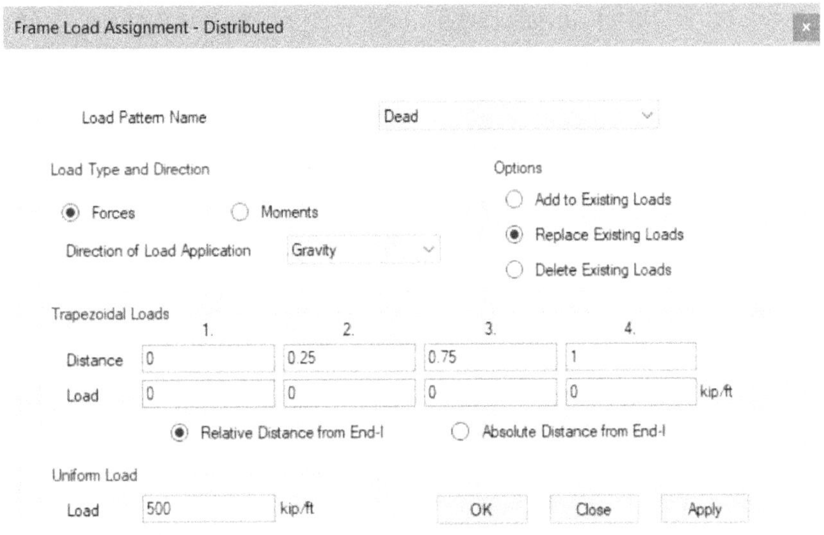

Figure-57. Frame Load Assignment-Distributed dialog box

- Specify the desired value of uniform load over selected frame members in the **Load** edit box of **Uniform Load** area of the dialog box.
- Set the other parameters as discussed earlier and click on the **OK** button.

Assigning Temperature to Selected Frame Members

The **Temperature** tool in the **Frame** cascading menu is used to assign temperature to selected frame members. The procedure to use this tool is given next.

- Select the frame members and click on the **Temperature** tool from the **Frame** cascading menu of the **Assign** menu. The **Frame Load Assignment-Temperature** dialog box will be displayed; refer to Figure-58.

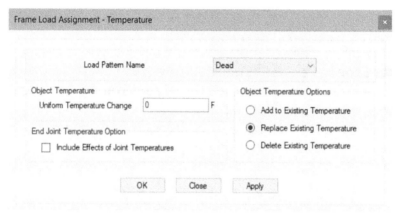

Figure-58. Frame Load Assignment-Temperature dialog box

- Specify the desired temperature value in the **Uniform Temperature Change** edit box.
- Select the **Include Effects of Joint Temperatures** check box to include the effect of temperatures applied to joints of the frame members.
- Set the other parameters as discussed earlier and click on the **OK** button.

Assigning Open Structure Wind to Frame Members

The **Open Structure Wind Parameters** tool in the **Frame Loads** cascading menu is used to assign wind parameters to selected frame members. The procedure to use this tool is given next.

- Click on the **Open Structure Wind Parameters** tool from the **Frame Loads** cascading menu of the **Assign** menu. The **Frame Load Assignment-Open Structure Wind Parameters** dialog box will be displayed; refer to Figure-59.
- Select the **Yes** option from the **Object is Loaded by Wind** drop-down to apply wind load on selected frame objects.
- Specify the desired value of ice thickness in the **Ice Thickness** edit box if you are designing frame for freezing temperature environment.

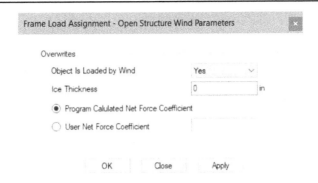

Figure-59. Frame Load Assignment-Open Structure
Wind Parameters dialog box

- Select the **Program Calculated Net Force Coefficient** radio button to apply default net force coefficient based on design code of your project.
- Select the **User Net Force Coefficient** radio button to specify the value of force coefficient manually.
- After setting the desired parameters, click on the **OK** button.

ASSIGNING SHELL LOADS

The tools in the **Shell Loads** cascading menu of the **Assign** menu are used to assign loads to selected shell objects; refer to Figure-60. The tools in this cascading menu are discussed next.

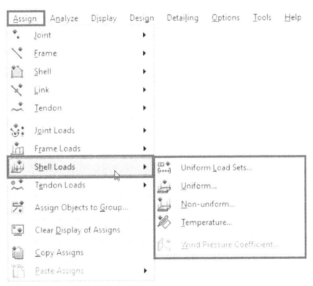

Figure-60. Shell Loads cascading menu

Assigning Uniform Load Sets

The **Uniform Load Sets** tool in the **Shell Loads** cascading menu is used to assign a combination of dead, live, post-tension transfer and post-tension final loads. The procedure to use this tool is given next.

- Select the desired shell objects and click on the **Uniform Load Sets** tool from the **Shell Loads** cascading menu of the **Assign** menu. The **Shell Assignment - Uniform Load Set** dialog box will be displayed; refer to Figure-61.

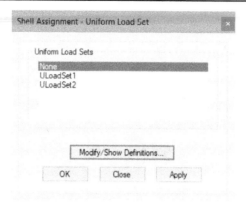

Figure-61. Shell Assignment-Uniform Load
Set dialog box

- Select the desired load set from the list and click on the **OK** button.

Defining Uniform Load Set

- If you have not created uniform load sets earlier then click on the **Modify/Show Definitions** button from the dialog box. The **Shell Uniform Load Sets** dialog box will be displayed; refer to Figure-62. Note that this dialog box can also be invoked by **Shell Uniform Load Sets** tool in the **Define** menu.
- Click on the **Add New Load Set** button from the dialog box to add a new load set. The **Shell Uniform Load Set Data** dialog box will be displayed; refer to Figure-63.

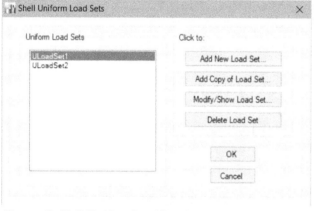

Figure-62. Shell Uniform Load Sets dialog box

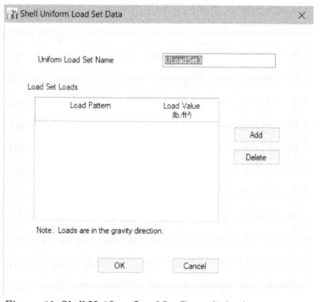

Figure-63. Shell Uniform Load Set Data dialog box

- Specify the desired name of load set in the **Uniform Load Set Name** edit box.
- Click on the **Add** button from the dialog box to add a load. A dead load of **0** value will be added. Click on the **Dead** option under **Load Pattern** column in the table and select the desired option like Live, Dead, PT-FINAL, and so on. Specify the value of load in adjacent edit box.

- Repeat the above step to add as many loads as you want in current load set. After adding desired load, click on the **OK** button.
- The **Shell Uniform Load Sets** dialog box will be displayed again with newly added load sets in the list. The other options in this dialog box are same as discussed in previous chapter.
- Click on the **OK** button from the dialog box. The **Shell Assignment - Uniform Load Set** dialog box will be displayed again. Select the desired load set and click on the **OK** button.

Assigning Uniform Shell Load

The **Uniform Shell Load** tool is used to assign a uniform load all over the selected shell objects. The procedure to use this tool is given next.

- Click on the **Uniform Shell Load** tool from the **Shell Loads** cascading menu of the **Assign** menu. The **Shell Load Assignment-Uniform** dialog box will be displayed; refer to Figure-64.

Figure-64. Shell Load Assignment-Uniform dialog box

- Select the desired type of load from the **Load Pattern Name** drop-down to define the type of load.
- Specify the desired value and direction of load in the **Uniform Load** area of the dialog box.
- Set the other parameters as discussed earlier and click on the **OK** button.

Assigning Non-uniform Load to Shell Objects

The **Non-uniform** tool is used to assign a varying load on selected shell objects. The procedure to use this tool is given next.

- Select the desired shell objects and click on the **Non-uniform** tool from the **Shell Loads** cascading menu of the **Assign** menu. The **Shell Load Assignment-Non-uniform** dialog box will be displayed; refer to Figure-65.

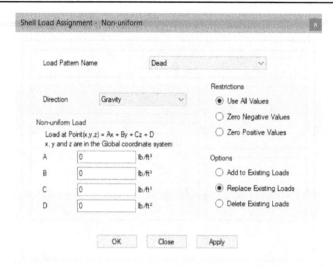

Figure-65. Shell Load Assignment-Non-uniform dialog box

- Specify the A, B, C, and D values of non-uniform load equation in respective edit boxes.
- Set the desired restriction from the **Restrictions** area of the dialog box. Select the **Use All Values** radio button to use all A, B, C, and D values specified. Select the **Zero Negative Values** radio button to use all positive values specified and set negative values 0. Select the **Zero Positive Values** radio button to use all negative values and set the positive values 0 in the formula.
- After setting the desired values, click on the **OK** button.

Applying Temperature to Shell Objects

The **Temperature** tool in the **Shell Loads** cascading menu is used to assign temperature to selected shell objects. The procedure to use this tool is same as discussed earlier.

ASSIGNING LOADS TO TENDONS

The tools to assign loads to tendons are available in the **Tendon Loads** cascading menu of the **Assign** menu; refer to Figure-66. These tools are discussed next.

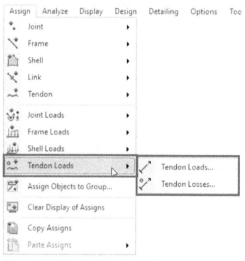

Figure-66. Tendon Loads cascading menu

Assigning Tendon Loads

The **Tendon Loads** tool is used to apply loads on selected tendons. The procedure to use this tool is given next.

* Select the tendons on which you want to apply the loads and click on the **Tendon Loads** tool from the **Tendon Loads** cascading menu of the **Assign** menu. The **Tendon Load** dialog box will be displayed; refer to Figure-67.

Figure-67. Tendon Load dialog box

* Select the desired options from the **Load Pattern Names** area. You will learn about creating load patterns in the next chapter.
* Specify the desired value in the **Tendon Jacking Stress** edit box to define jacking stress.
* From the **Jack From This Location** area select the radio button to define the location at which jacking stress will be applied.
* After setting the desired parameters, click on the **OK** button.

Assigning Tendon Losses

The **Tendon Losses** tool is used to apply losses to selected tendons. The procedure to use this tool is given next.

* Select the tendons and click on the **Tendon Losses** tool from the **Tendon Loads** cascading menu in the **Assign** menu. The **Tendon Loss Options** dialog box will be displayed; refer to Figure-68.

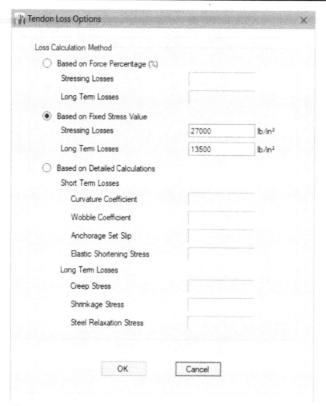

Figure-68. Tendon Loss Options dialog box

- Select the desired radio button to select loss calculation method. Select the **Based on Force Percentage (%)** radio button to specify the percentage of force that will be lost after post tensioning. Select the **Based on Fixed Stress Value** radio button to specify fix value of load will be lost after tensioning tendons. Select the **Based on Detailed Calculations** radio button to specify various coefficients for short term losses and values for long term losses.
- After selecting the desired radio button, specify the related values in respective edit boxes.
- Click on the **OK** button. The specified losses will be applied to selected tendon objects.

COPYING AND PASTING ASSIGNS

The **Copy Assigns** tool in the **Assign** menu is used to copy assigned properties and parameters of selected objects so that you can apply the same to other similar objects. The **Paste Assigns** tools in the **Assign** menu are used to assign copied properties to selected objects. The procedure to copy and past assigns is discussed next.

- Select the object whose assigns are to be copied and click on the **Copy Assigns** tool from the **Assign** menu; refer to Figure-69.

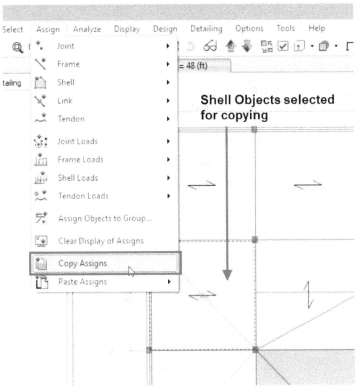

Figure-69. Copying assigns

- Now, select the same type of objects to which you want to assign copied properties and click on the respective button from the **Paste Assigns** cascading menu. For example, to paste assigns copied from a shell object, select the **Shell** tool from the **Paste** cascading menu and apply them to another shell objects.

FOR STUDENT NOTES

Chapter 5

Creating Load Cases and Performing Analysis

Topics Covered

The major topics covered in this chapter are:

- *Introduction*
- *Defining Load Patterns*
- *Creating Modal Cases*
- *Setting Load Cases*
- *Setting Load Combinations*
- *Creating Auto Construction Sequence Load Case*
- *Creating Walking Vibrations*
- *Checking Model for Analysis*
- *Setting Degree of Freedom for Analysis*
- *Setting Load Cases to Run Analysis*

- *Setting SAPFire Solver Options*
- *Mesh Setting for Floors and Walls*
- *Setting Hinges and Cracking Analysis*
- *Running Analysis*
- *Applying Modal Alive*
- *Modifying Undeformed Geometry*
- *Checking Analysis Log*
- *Unlocking Model*
- *Setting Display Options*

INTRODUCTION

In previous chapters, we have learned to assign properties and loads to various frame members. In this chapter, we will learn to set load cases and combinations for analysis. You will also learn to perform analysis under specified load conditions. Later in this chapter, you will learn to analyze the results of analysis.

DEFINING LOAD PATTERNS

The **Load Pattern** tool is used to define the distribution of automatic loads like self weight, wind, earth quake and other phenomenons for design studies. You will learn more about design studies in next chapter. Note that Load pattern is the first step for applying load in the model. After creating load pattern, you will assign different loads in the model using the **Assign** menu options. These assigned loads will fall under different categories of load pattern. Once you have assigned different loads in respective patterns then you will create load cases to run analysis. The procedure to create load pattern is given next.

- Click on the **Load Pattern** tool from the **Define** menu. The **Define Load Patterns** dialog box will be displayed; refer to Figure-1.

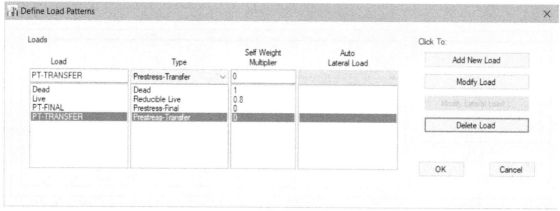

Figure-1. Define Load Patterns dialog box

- Specify the desired name of load in the **Load** edit box.
- Click in the **Type** drop-down and select the desired type of load. Like to apply wind load, select the **Wind** option from the **Type** drop-down.
- Specify the desired value in the **Self-weight Multiplier** edit box to define the amount of weight to be added to the current load.
- Set the desired auto lateral loading code in the **Auto Lateral Load** drop-down if you are defining seismic or wind load.
- Click on the **Add New Load** button to add it in the pattern.
- To modify a load, select it from the list and change the parameters as required from the top of the table. After setting desired changes, click on the **Modify Load** button.
- If you want to modify lateral loading of a seismic or wind load then select it from the table and click on the **Modify Lateral Load** button. The **Wind Load Pattern** or **Seismic Loading** dialog box will be displayed based on the type of load; refer to Figure-2 and Figure-3.

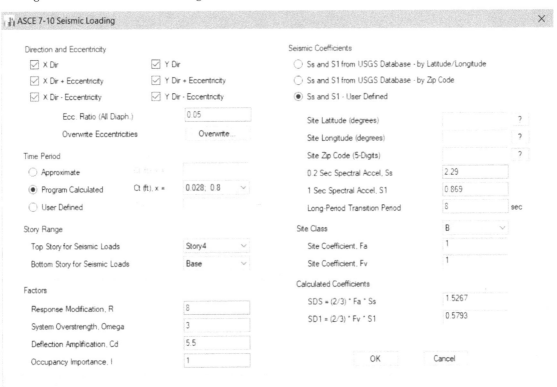

Figure-2. Wind Load Pattern dialog box

Figure-3. Seismic Loading dialog box

Modifying Lateral load of wind

• Select the **Exposure from Extents of Diaphragms** radio button to specify wind load individually to each of the diaphragms. A separate lateral load is created for each diaphragm present at a story level. The wind loads calculated at any story level are based on the story level elevation, the story height above and below that level, the assumed exposure width for the diaphragm(s) at the story level, and various code-dependent wind coefficients.

- Select the **Exposure from Frame and Shell Objects** radio button to specify wind load for each frame and shell object as per the wind pressure coefficient, C_p. Select the **Include Shell Objects** check box to include shell objects for wind loading. Select the **Include Frame Objects (Open Structure)** check box to include open frames for wind loading.

Wind Pressure Coefficients

- Specify the desired values of wind pressure coefficients in the **Wind Pressure Coefficients** area of the dialog box if you have selected the **Exposure from Extents of Diaphragms** radio button in the dialog box. Note that the edit boxes in the **Wind Pressure Coefficients** area will be active only if you have **User Specified** radio button from the **Wind Pressure Coefficients** area. Select the **Program Determined** radio button if you want the wind pressure coefficients to be determined automatically based on selected design codes.

Wind Exposure Parameters

The all options in this area of the dialog box are active only when the **Exposure from Extents of Diaphragms** radio button is selected in the dialog box.

- Click on the **Modify/Show** button from the **Wind Exposure Parameters** area to set wind direction and width. The **Wind Exposure Width Data** dialog box will be displayed; refer to Figure-4.

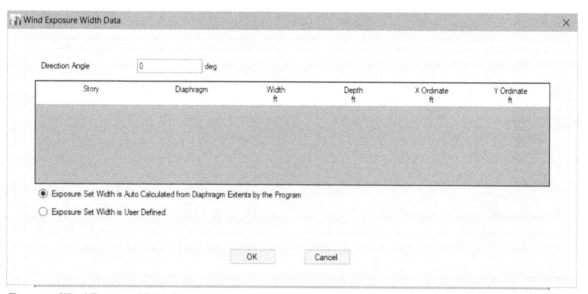

Figure-4. Wind Exposure Width Data dialog box

- Select the **Exposure Set Width is Auto Calculated from Diaphragm Extents by the Program** radio button to automatically set the width of wind exposure automatically. Specify the angle value in the **Direction Angle** edit box.
- Select the **Exposure Set Width is User Defined** radio button to also set the width of exposure manually. To add wind exposure, click on the **Add Row** button. A new row will be added in the list. Set the desired parameters in the table.

- Select the desired case from the **Case** drop-down. Hover the cursor on information button (ℹ️)to check the details of cases in the drop down.
- Set the desired values of **e1 Ratio** and **e2 Ratio** in respective edit boxes.

Wind Coefficients

The options in the **Wind Coefficients** area is used to specify basic parameters of wind.

- Specify the desired value of wind speed in miles per hour in the **Wind Speed (mph)** edit box.
- Select the desired option from the **Exposure Type** drop-down. The **B** type is used for urban and suburban areas, wooded areas, or other terrain with numerous closely spaced obstructions having the size of single-family dwellings or larger. The **C** type is used for open terrain with scattered obstructions having heights generally less than 30 feet. (Commonly associated with flat open country and grasslands). Select the **D** type for structures at a close distance (typically within 600 feet) from an "open waterway" one mile or more across. This category is readily distinguishable, where the locally enforced code very likely has considered this in their requirements.
- Set the desired value in **Topographical Factor, Kzt** edit box. There are different values of Topographical factor for different type of terrains. Like for plain terrain value of Kzt is 1. For other terrains, height of ridges, escarpments, and hills are considered to derive the factor.
- Specify the desired value of Gust factor in the **Gust Factor** edit box. A gust factor, defined as the ratio between a peak wind gust and mean wind speed over a period of time, can be used along with other statistics to examine the structure of the wind.
- Specify the desired value of directionality factor in the **Directionality Factor, Kd** edit box. The directionality factor (Kd) used in the ASCE 7 wind load provisions for components and cladding is a load reduction factor intended to take into account the less than 100% probability that the design event wind direction aligns with the worst case building aerodynamics. the Directionality Factor Kd is defined as a parameter that makes the design more rational by considering the dependencies of the wind speed, the frequency of occurrence of extreme wind and the aerodynamic property on wind direction. The wind directionality factor Kd is affected by the frequency of occurrence and the routes of typhoons, climatological factors, large-scale topographic effects, and so on.

Exposure Height

- Set the desired upper and lower limit in the **Top Story** and **Bottom Story** edit boxes respectively in the **Exposure Height** area of the dialog box.

- If you have selected the **Exposure from Extents of Diaphragms** radio button then the **Include Parapet** check box will be active. Select the check box to include parapets while considering wind loads. A parapet is a barrier which is an extension of the wall at the edge of a roof, terrace, balcony, walkway or other structure. Specify the desired value of height (ft) in the **Parapet Height** edit box.

- After setting desired parameters, click on the **OK** button to apply modifications to lateral loads.

You can modify seismic loading in the same way.

CREATING MODAL CASES

The modal case is used to find natural frequencies of the model. The procedure to create modal case is given next.

- Click on the **Modal Cases** tool from the **Define** menu. The **Modal Cases** dialog box will be displayed with list of modal cases earlier created; refer to Figure-5.

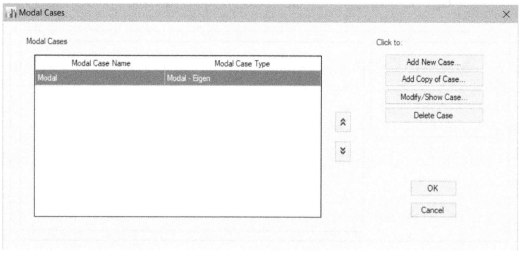

Figure-5. Modal Cases dialog box

- Click on the **Add New Case** button. The **Modal Case Data** dialog box will be displayed; refer to Figure-6.
- Specify the desired name in the **Modal Case Name** edit box.
- Select the desired option from the **Modal Case Sub Type** drop-down. For dynamic analysis, Ritz vectors are recommended over Eigen vectors because, for the same number of modes, Ritz vectors provide a better participation factor, which enables the analysis to run faster with the same level of accuracy. When analysis involves ground motion in the horizontal plane, the benefit is not as pronounced. However, for vertical acceleration, localized machine vibration, and the nonlinear FNA method, Ritz vectors are much more well-suited for analysis.

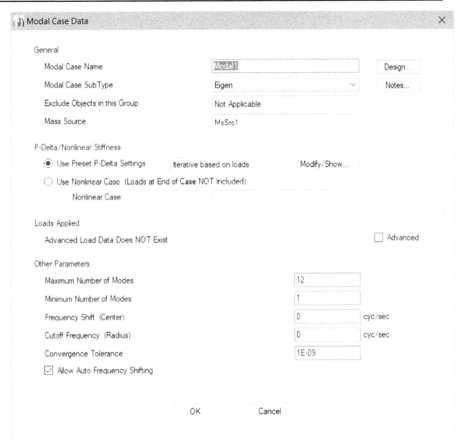

Figure-6. Modal Case Data dialog box

- If you have selected **Eigen** option from the drop-down then select the **Advanced** check box to define loads along with modal analysis. The options in the **Loads Applied** area will be displayed as shown in Figure-7. Specify the loads as discussed earlier. Similarly, you can add load for Ritz load case sub-type.

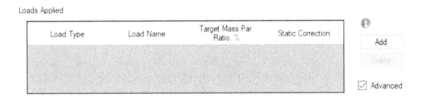

Figure-7. Loads Applied area with Eigen load case subtype

- If you have selected Ritz Modal Case SubType then specify the maximum and minimum modes in **Maximum Number of Modes** and **Minimum Number of Modes** edit boxes, respectively.
- Specify the desired values for frequency shift, cutoff frequency and convergence tolerance in respective edit boxes if you have selected **Eigen** option in **Modal Case SubType** drop-down and click on the **OK** button.
- Note that if you want to define a design load case then click on the **Design** button in the dialog box and specify the parameters.

SETTING LOAD CASES

A load case defines how load patterns are applied (statically or dynamically), how the structure responds (linearly or nonlinearly), and how analysis is performed (through modal analysis, direct integration, etc.). For each analysis to be performed, a load case is defined. Each load case may apply a single load pattern or a combination of load patterns. The procedure to create and manage load cases is given next.

- Click on the **Load Cases** tool from the **Define** menu. The **Load Cases** dialog box will be displayed with a list of load patterns earlier created; refer to Figure-8.

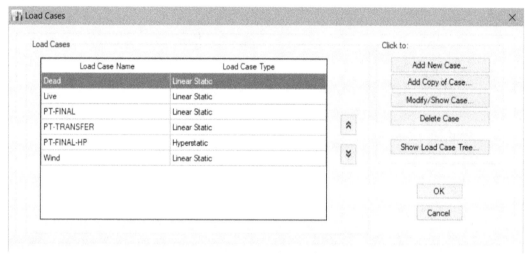

Figure-8. Load Cases dialog box

- To add a new load case, click on the **Add New Case** button. The **Load Case Data** dialog box will be displayed; refer to Figure-9.
- Set the desired name for load case in the **Load Case Name** edit box.
- Select the desired load case type from the **Load Case Type** drop-down. Select the **Linear Static** option if you want to apply fix loads without any dynamic effects. Select the **Nonlinear Static** option if you want to apply time changing force and check the large displacement effects in the model. Select the **Nonlinear Staged Construction** option to perform analysis at different stages. Select the **Response Spectrum** option to check the plot of the peak or steady-state response (displacement, velocity or acceleration) of a series of oscillators of varying natural frequency, that are forced into motion by the same base vibration or shock. Select the **Time History** option from the drop-down if Time-varying loads are applied. The solution may be by modal superposition or direct integration methods. Select the **Buckling** option from the drop-down if you are concerned about the buckling failure of the structure. Select the **Hyperstatic** option to apply hyperstatic load. A hyperstatic load case is required in certain codes when performing design with post-tensioning.

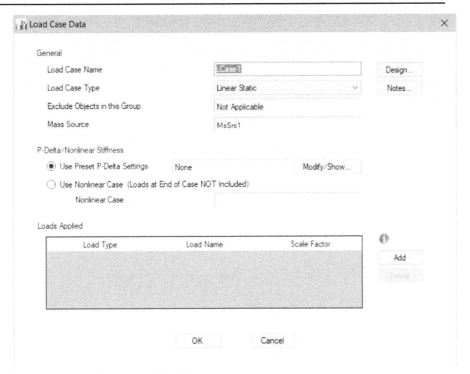

Figure-9. Load Case Data dialog box

- To move the load cases up or down, select the load case from the list and click on the **Up** or **Down** button at the right in the dialog box. Note that moving the load cases up or down does not define their position while performing analysis. All the load cases will run as per their dependency.
- The options in this dialog box will be modified accordingly. Set the parameters as required and click on the **OK** button. You will learn more about these options later.
- The **Add Copy of Case**, **Modify/Show Case**, and **Delete Case** buttons in the dialog box work in the same way as they are explained in previous chapters.
- Click on the **OK** button from the **Load Cases** dialog box apply the load cases.

Creating Linear Static Load Case

The Linear Static load case is created to perform analysis with static load on the model. The procedure to create linear static load case is given next.

- Click on the **Add New Case** button from the **Load Cases** dialog box. The **Load Case Data** dialog box will be displayed as shown in Figure-9.
- Specify the desired name for the case in the **Load Case Name** edit box.
- Select the **Linear Static** option from the **Load Case Type** drop-down to create linear static case.
- Select the **Use Preset P-Delta Settings** radio button to apply P-Delta settings to the current linear static analysis. For most tall building structures, the P-Delta effect is of most concern in the

columns because of gravity load, including dead and live load.

- Click on the **Modify/Show** button to specify parameters for P-Delta. The **Preset P-Delta Options** dialog box will be displayed; refer to Figure-10.

Figure-10. Preset P-Delta Options dialog box

- Select the **Non-iterative-Based on Mass** radio button to automatically compute load at each level as a story-by-story load on the structure. Note that this approach is approximate and do not require iterative solution. This method essentially treats the building as a simplified stick model to consider the P-Delta effect. It does not capture local buckling as well as the iterative method. This method works best if the model has a single diaphragm at each floor level, although it also works for other cases as well. Select the **Iterative - Based on Loads** radio button to compute load cases by specified combination of static load patterns. For example, the load case may be the sum of a dead load case plus a fraction of a live load case. This approach requires an iterative solution to determine the P-Delta effect upon the structure. This method considers the P-Delta effect on an element-by-element basis. It captures local buckling effects better than the non-iterative method. We recommend the use of this iterative method in all cases except those for which no gravity load is specified in the model. On selecting this radio button, the options in the **Iterative P-Delta Load Case** area will become active. Select the desired load pattern and respective scale factor in the table and click on the **Add** button. Click on the **OK** button to set preset of P-Delta.
- Select the **Use Nonlinear Case (Loads at Ends of Case NOT Included)** radio button to set nonlinear stiffness for the analysis. The drop-down below the radio button will become active. Select the desired nonlinear case earlier created from the **Nonlinear Case** drop-down.
- Click in the field under **Load Type** column and select the desired load type. Select **Load Pattern** option to add desired load. Select the **Acceleration** option to apply acceleration load on the model. Note that Static, modal, and buckling load cases permit only acceleration

loads, UX, UY, UZ, RX, RY, and RZ. Response-spectrum and time-history load cases permit application of ground acceleration in a local coordinate system. Specify an angle of rotation about the Z axis. The acceleration load names are U1, U2, U3, R1, R2, and R3.
- Click in the field under **Load Name** column and select the desired load.
- Specify the desired scale factor in the **Scale Factor** column. Click on the **Add** button to add the current load in load case data. Add as many loads as desired and click on the **OK** button from the dialog box to create load case.

Creating Nonlinear Static Load Case

The Nonlinear Static load case is created to perform analysis with specified initial conditions and static load on the model. The procedure to create Nonlinear static load case is given next.

- Click on the **Add New Case** button from the **Load Cases** dialog box. The **Load Case Data** dialog box will be displayed. Select the **Nonlinear Static** option from the **Load Case Type** drop-down. The dialog box will be displayed as shown in Figure-11.

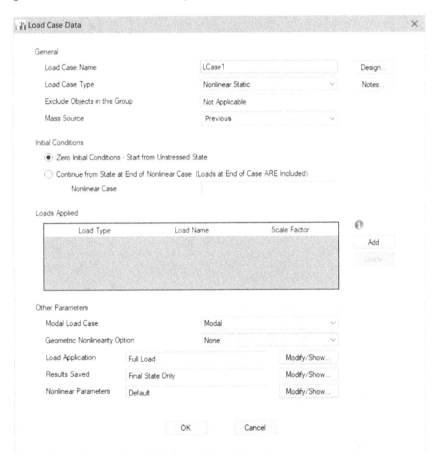

Figure-11. Load Case Data dialog box for Nonlinear Static case

- Select the desired mass source from the **Mass Source** drop-down.
- Select the **Zero Initial Conditions - Start from Unstressed State** radio button to set zero displacement and velocity. All elements are unstressed and there is no history of nonlinear deformation.

- Select the **Continue from State at End of Nonlinear Case (Loads at End of Case ARE Included)** radio button to select desired nonlinear load whose end results will be used as starting point for current analysis. Select the desired load case from the **Nonlinear Case** drop-down.
- Add the desired loads in the table in **Loads Applied** area as discussed earlier.
- Select the desired option from the **Modal Load Case** drop-down. All the modal load cases created earlier are displayed in this drop-down.
- Select the desired option to define non-linearity of geometry in the **Geometric Nonlinearity Option** drop-down. If you have selected **None** option then the All equilibrium equations are considered in the undeformed configuration of the structure. If you have selected **P-Delta** option then The equilibrium equations take into partial account the deformed configuration of the structure. Tensile forces tend to resist the rotation of elements and stiffen the structure, and compressive forces tend to enhance the rotation of elements and destabilize the structure. This may require a moderate amount of iteration. If you have selected **P-Delta plus Large Displacements** option then all equilibrium equations are written in the deformed configuration of the structure. This may require a large amount of iteration. Although large displacement and large rotation effects are modeled, all strains are assumed to be small.

Defining Load Application Parameters

- Click on the **Modify/Show** button next to **Load Application** field in the dialog box to specify how load will be applied. The **Load Application Control for Nonlinear Static Analysis** dialog box will be displayed; refer to Figure-12.

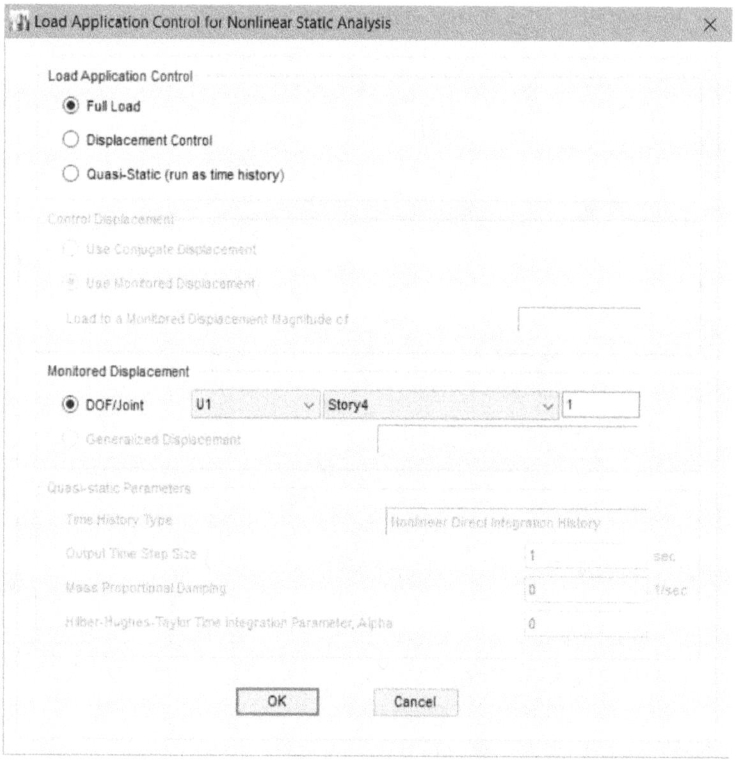

Figure-12. Load Application Control for Nonlinear Static Analysis dialog box

- Select the **Full Load** radio button if you want to apply full load at once. Select the **Displacement Control** radio button if you know the distance the structure is to move is known, but the amount of load is unknown. This is most useful for structures that become unstable and may lose load-carrying capacity during the course of the analysis. Typical applications include static pushover analysis and snap-through buckling analysis. Select the **Quasi-Static** radio button when the distance the structure moves is known. However, the solution is found by applying a specified amount of load incrementally, and solving using dynamic analysis rather than static analysis. This is most useful for structures that become unstable and may lose load-carrying capacity during the course of the analysis, and for which a solution could not be found using displacement control. Typical applications include static pushover analysis and snap-through buckling analysis. Set the other parameters as required and click on the **OK** button.

- Click on the **Modify/Show** button next to **Results Saved** field. The **Results Saved for Nonlinear Static Case** dialog box will be displayed; refer to Figure-13. Select the **Final State Only** radio button to save the result of analysis only once and that too at the end of analysis. Select the **Multiple States** radio button to save the results of non-linear analysis at regular intervals as specified in the dialog box. Click on the **OK** button to apply the settings.

Figure-13. Results Saved for Nonlinear Static Case dialog box

- Click on the **Modify/Show** button next to **Nonlinear Parameters** field. The **Nonlinear Parameters** dialog box will be displayed; refer to Figure-14. Set the desired parameters and click on the **OK** button.

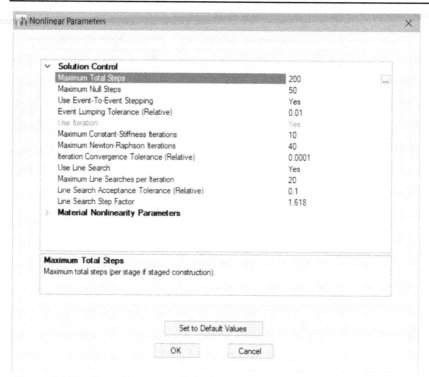

Figure-14. Nonlinear Parameters dialog box

- After setting desired parameters, click on the **OK** button to create load case.

Creating Nonlinear Staged Construction Load Case

The Nonlinear staged construction load case is created to perform non-linear analysis at different stages. The procedure to create this load case is given next.

- Click on the **Add New Case** button from the **Load Cases** dialog box and select the **Nonlinear Staged Construction** option from the **Load Case Type** drop-down. The **Load Case Data** dialog box will be displayed as shown in Figure-15.
- Most of the parameters are same as discussed earlier.
- Click on the **Add** button in the **Stage Definition** area to create a new stage at which analysis will be performed. A new row will be added in the table. Double-click in the field of table to change the parameters.
- Select the desired stage and click on the **Stage Operations** button to add operations. The **Stage Data** dialog box will be displayed; refer to Figure-16.

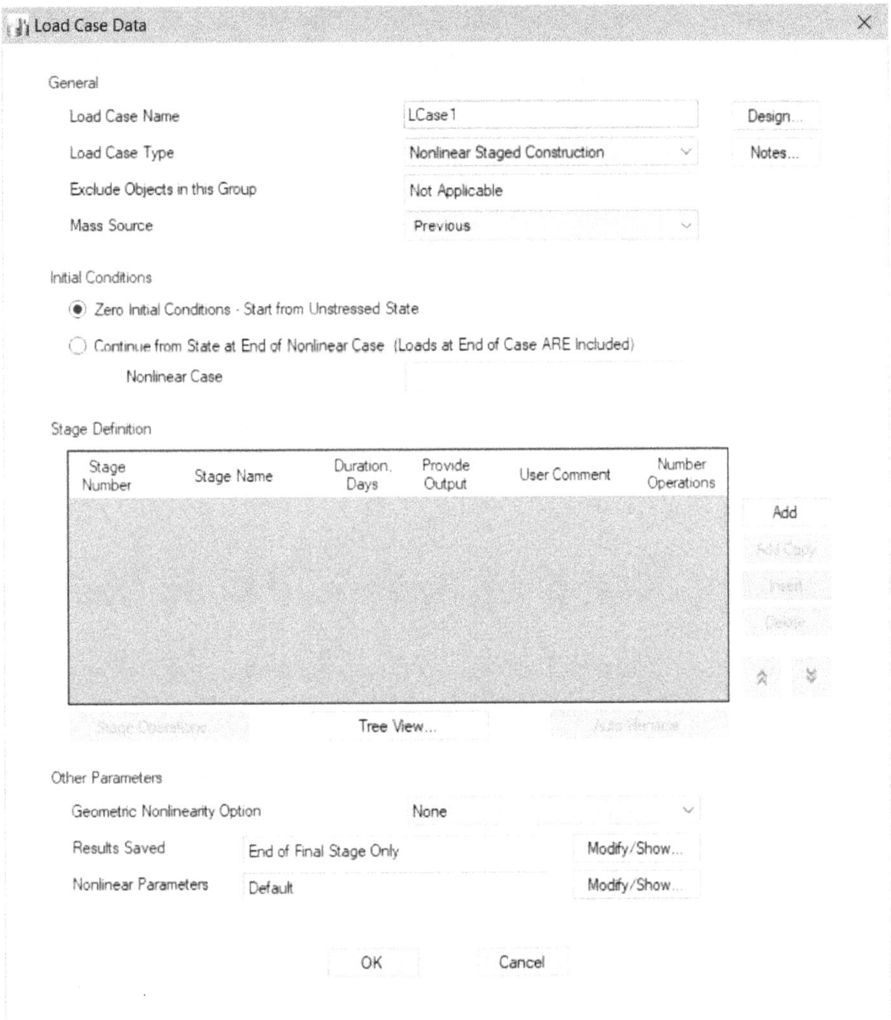

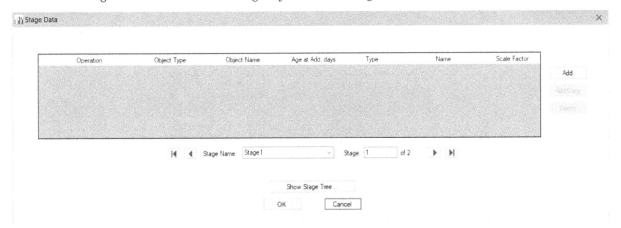

Figure-15. Load Case Data dialog box for Nonlinear Staged Construction case

Figure-16. Stage Data dialog box

- Select the desired stage from the **Stage Name** drop-down. Click on the **Add** button. A new operation will be added in the list.
- Click in the field under **Operation** column and select the desired type of operation like Adding structure, removing structure, loading objects, and so on.
- Click in the field under **Object Type** column and select the desired type of object on which you want to perform the selected operation.

- Specify the age of structure being added in current stage in the **Age at Add, days** edit box. This column is active only when **Add Structure** option selected in the **Operation** column.
- If you have selected the **Load Objects** or **Load Objects if Added** in the **Operation** column then **Type**, **Name**, and **Scale Factor** columns will be active. Specify the desired load type, load name and scale factor under respective columns.
- If you have selected the **Change Sections** option under **Operation** column then specify the section type and size in the **Type** and **Name** columns respectively.
- After setting desired data, click on the **OK** button. The **Load Case Data** dialog box will be displayed again.
- Set the other parameters as discussed earlier and click on the **OK** button to create the load case.

Creating Response Spectrum Load Case

The response spectrum load case is used to check the response of structure under vibrational loads. The procedure to create this load case is given next.

- Click on the **Add New Case** button from the **Load Cases** dialog box and select the **Response Spectrum** option from the **Load Case Type** drop-down. The **Load Case Data** dialog box will be displayed as shown in Figure-17.

Figure-17. Load Case Data dialog box for Response Spectrum case

ETABS 2016 Black Book   5-17

- Select the desired modal combination method from the **Modal Combination Method** drop-down. The options in this drop-down are as below:

CQC - Complete Quadratic Combination option. A modal combination technique that accounts for modal damping. It is the same as SRSS if damping is zero. Enter the characteristic frequencies f1 and f2 as defined in ASCE 4 for GMC.

SRSS - Square Root of Sum of Squares option. A modal combination technique that does not account for modal damping or cross coupling. Enter the characteristic frequencies f1 and f2 as defined in ASCE 4 for GMC.

Absolute option. Summation of the absolute values of the modal results.

GMC - General Modal Combination option. A modal combination technique that takes into account modal damping, and assumes higher correlation between modes at higher frequencies. Enter the characteristic frequencies f1 and f2 as defined in ASCE 4 for GMC.

NRC 10 Pct - the Ten Percent method of the U.S. Nuclear Regulatory Commission Regulatory Guide 1.92 option. The Ten Percent method assumes full, positive coupling between all modes whose frequencies differ from each other by 10% or less of the smaller of the two frequencies. Modal damping does not affect the coupling. Enter the characteristic frequencies f1 and f2 as defined in ASCE 4 for GMC.

Double Sum - the Double Sum method of the U.S. Nuclear Regulatory Commission Regulatory Guide 1.92 option. The Double Sum method assumes a positive coupling between all modes, with correlation coefficients that depend upon damping in a fashion similar to the CQC and GMC methods, and that also depend upon the duration of the earthquake. Specify this duration as parameter td as part of the load cases definition.

CQC3 - An extension of the SRSS method for finding the maximum response when the horizontal (U1 and U2) directions of loading use the same response spectrum function but have different scale factors. The critical angle of loading is determined automatically independent of the angle specified for the loading. The vertical response is combined with the maximum horizontal response using the SRSS method. If different response-spectrum functions are used for U1 and U2, the results must be interpreted carefully by the engineer.

- Select the **Include Rigid Response** check box to include rigid frequencies. Specify the desired parameters in edit boxes adjacent to the check box.
- Set the modal damping and diaphragm eccentricity values using the **Modify/Show** buttons.
- After setting the desired values, click on the **OK** button.

Creating Time History Load Case

The Time History load case is used when time varying loads are applied. The procedure to use this tool is given next.

- Click on the **Add New Case** button from the **Load Cases** dialog box and select the **Response Spectrum** option from the **Load Case Type** drop-down. The **Load Case Data** dialog box will be displayed as shown in Figure-18.

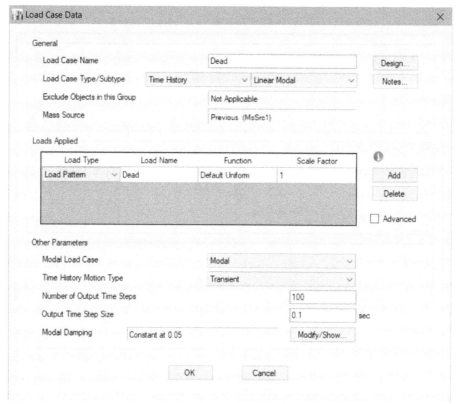

Figure-18. Load Case Data dialog box for Time History case

- Select the desired subtype of Time History load case from the **Subtype** drop-down. The options will be modified in the dialog box accordingly.

Setting Linear Modal Time History Load Case

- Select the **Time History** option from the **Load Case Type** drop-down and select the **Linear Modal** option from the **Subtype** drop-down. The options will be displayed as shown in Figure-18.
- Set the loads applied and Modal Load Case as discussed earlier.
- Select the **Transient** option from the **Time History Motion Type** drop-down when structure starts at rest and is subjected to the specified loads only during the time period specified for the analysis. The Select the **Periodic** option from the drop-down when specified loads are assumed to be periodic i.e. they repeat indefinitely with a period given by the length of the analysis.
- Specify the desired number of steps in the **Number of Output Time Steps** edit box.
- Set the desired value of time in the **Output Time Step Size** edit box.
- After setting desired parameters, click on the **OK** button.

Setting Nonlinear Modal Time History Load Case

- Select the **Time History** option from the **Load Case Type** drop-down and select the **Nonlinear Modal (FNA)** option from the **Subtype** drop-down. The options will be displayed as shown in Figure-19.

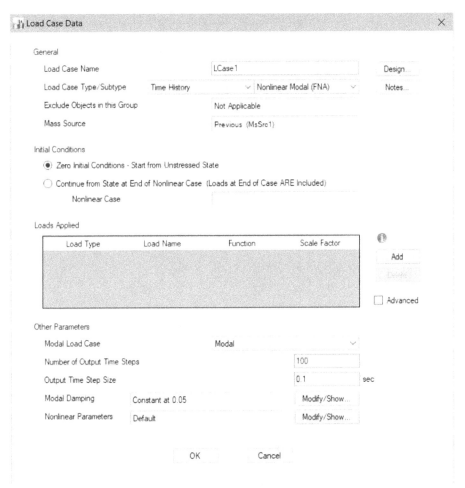

Figure-19. Load Case Data dialog box for Time History Nonlinear Modal case

- Set all the parameters as discussed earlier and click on the **OK** button to create load case.

Setting Linear Direct Integration Time History Load Case

- Select the **Time History** option from the **Load Case Type** drop-down and select the **Linear Direct Integration** option from the **Subtype** drop-down. The options will be displayed as shown in Figure-20.
- Click on the **Modify/Show** button next to **Damping** field in the dialog box. The **Direct Integration Damping** dialog box will be displayed; refer to Figure-21. The options in this dialog box are used to specify the method by which damping will be applied.

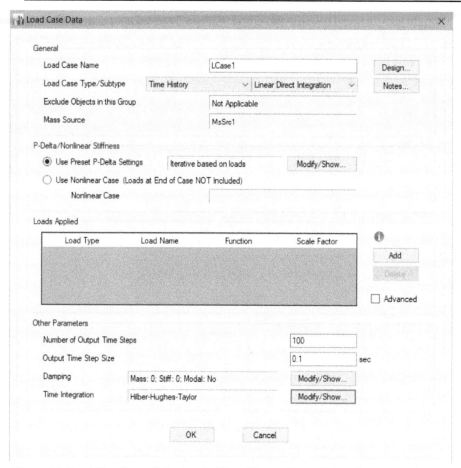

Figure-20. Load Case Data dialog box for Time History Linear Direct Integration load case

Setting Direct Integration Damping

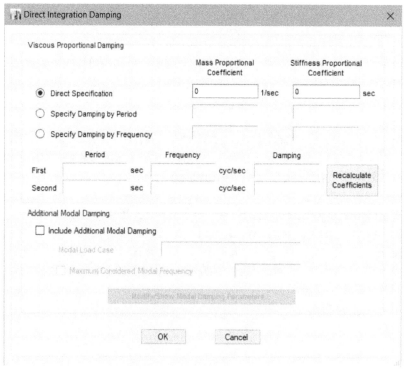

Figure-21. Direct Integration Damping dialog box

- Select the **Direct Specification** radio button to specify the mass and stiffness coefficient of damping.

- Select the **Specify Damping by Period** radio button to specify the time period and relative damping in the table. After specifying the time period and damping value, click on the **Recalculate Coefficient** button.
- If you want to add modal damping also to the damping coefficient then select the **Include Additional Modal Damping** check box. The options below it will become active. Set the desired parameters.
- Select the **Specify Damping by Frequency** radio button to specify frequency and damping in the table. After setting parameters, click on the **Recalculate Coefficients** button.
- Click on the **OK** button to apply direct integration damping.

Specifying Time Integration Method

The Time Integration methods used to define how the time integration will be performed.

- Click on the **Modify/Show** button next to Time Integration field. The **Time Integration Parameters** dialog box will be displayed; refer to Figure-22.

Figure-22. Time Integration Parameters dialog box

- Select the desired radio button to specify time integration method. Set the respective parameters and click on the **OK** button. The **Load Case Data** dialog box will be displayed again.
- Set the other parameters in the **Load Case Data** dialog box as discussed earlier.

Setting Nonlinear Direct Integration Time History Load Case

- Select the **Time History** option from the **Load Case Type** drop-down and select the **Nonlinear Direct Integration** option from the **Subtype** drop-down. The options will be displayed as shown in Figure-23.

Figure-23. Load Case Data dialog box for Time History Nonlinear Direct Integration load
case

- All the options in this dialog box have been discussed earlier. Set the desired parameters and click on the **OK** button to create load case.

Creating Buckling Load Case

The Buckling load case is used when sudden failure of structure occurs due to buckling. The procedure to create this case is given next.

- Click on the **Add New Case** button from the **Load Cases** dialog box and select the **Response Spectrum** option from the **Load Case Type** drop-down. The **Load Case Data** dialog box will be displayed as shown in Figure-24.
- Specify the desired number of buckling modes to be found in the **Number of Buckling Modes** edit box.
- Specify the desired value of convergence tolerance in the **Eigenvalue Convergence Tolerance** edit box.
- Set the other parameters in the dialog box as discussed earlier.
- After setting the parameters, click on the **OK** button.

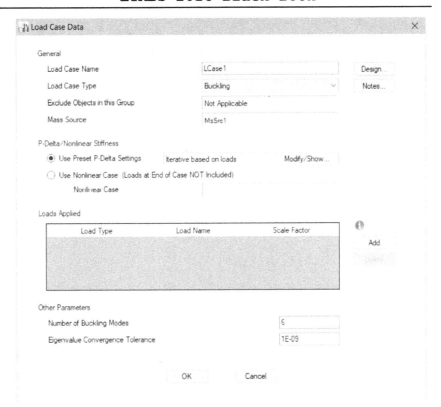

Figure-24. Load Case Data dialog box for Buckling load case

Creating Hyperstatic Load Case

A hyperstatic load case calculates the linear response of the structure with all supports removed and loaded only by the reactions from another linear static load case. This is typically used to calculate the secondary forces under prestress loading. The procedure to create this case is given next.

- Click on the **Add New Case** button from the **Load Cases** dialog box and select the **Hyperstatic** option from the **Load Case Type** drop-down. The **Load Case Data** dialog box will be displayed as shown in Figure-25.

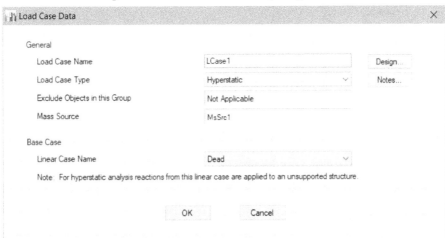

Figure-25. Load Case Data dialog box for Hyperstatic load case

- Select the desired option from the **Linear Case Name** drop-down to set the load whose reaction will be used for analysis.
- After setting the desired parameters, click on the **OK** button to create load case.

SETTING LOAD COMBINATIONS

The load combinations are used by ETABS to perform automatic designing of structure. In ETABS, load combinations, or combos, are generated automatically by the program or are user defined. If the automatically generated load combinations are acceptable (note that these combos are created/recreated after each design run), no definition of additional load combinations is required. The procedure to create and edit load combinations is given next.

- Click on the **Load Combinations** tool from the **Define** menu. The **Load Combinations** dialog box will be displayed; refer to Figure-26. Some of design load combinations created automatically by ETABS are displayed in the **Combinations** list box.

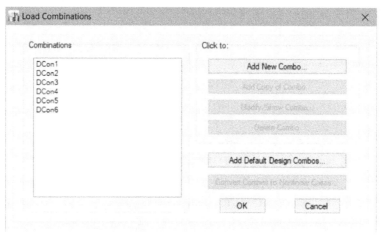

Figure-26. Load Combinations dialog box

Adding New Combination

- Click on the **Add New Combo** button from the dialog box. The **Load Combination Data** dialog box will be displayed; refer to Figure-27.

Figure-27. Load Combination Data dialog box

- Specify the desired name in the **Load Combination Name** edit box will be displayed.
- Select the desired combination from the **Combination Type** drop-down. Select the **Linear Add** option to perform algebraic addition of loads. Select the **Envelope** option to set combo maximum the maximum of all of the maximum values for each of the included analysis cases. Similarly, the combo minimum is the minimum of all of the minimum values for each of the included analysis cases. Select the **Absolute Add** option to set combo maximum, the sum of the larger absolute values for each of the included analysis cases. The combo minimum is the negative of the combo maximum. Select the **SRSS** option to set combo maximum, the square root of the sum of the squares of the larger absolute values for each of the included analysis cases. The combo minimum is the negative of the combo maximum. Select the **Range Add** option to set combination maximum, the sum of the positive maximum values from each of the contributing cases (a case with a negative maximum does not contribute), and the combined minimum, the sum of the negative minimum values from each of the contributing cases (a case with a positive minimum does not contribute). This combo type is useful for pattern or skip-type loading where all permutations of the contributing load case must be considered.
- Click on the **Modify/Show Notes** button to add or display notes to the load combination.
- Select the desired load pattern from the **Load Name** column in the table for first load; refer to Figure-28. Specify the desired factor value in the **Scale Factor** column.

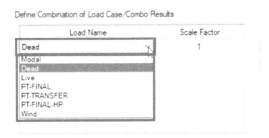

Figure-28. Selecting load

- Click on the **Add** button from the right of the table to add a new load and modify its parameters as discussed earlier.
- Set the other parameters as required and click on the **OK** button. The combination will be added in the **Combinations** list box.

Adding Default Design Combinations in the List

The procedure to add default design combinations is given next.

- Click on the **Add Default Design Combos** button from the dialog box to add default design combinations. The **Add Default Design Combinations** dialog box will be displayed; refer to Figure-29.
- Select the desired check box to include load combinations from the different design codes. Like, select the **Steel Frame Design** check box to include default load combinations of steel frame design in the list.

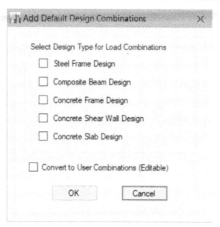

Figure-29. Add Default Design Combi-
nations dialog box

- Select the **Covert to User Combinations (Editable)** check box to convert the default load combinations to user combinations so that they can be edited.
- After setting the desired parameters, click on the **OK** button.

Converting Combination to Nonlinear Case

The **Convert Combos to Nonlinear Cases** button is used to convert selected load combination into a nonlinear case. Select the desired combination and click on the **Convert Combos to Nonlinear Cases** button to set it as a nonlinear case.

CREATING AUTO CONSTRUCTION SEQUENCE LOAD CASE

An auto sequential construction case provides an automated method of creating static nonlinear staged construction cases that are specifically tailored to model construction sequence loading. Note that only one auto sequential construction case can be defined per model file. The procedure to create auto construction sequence load case is given next.

- Click on the **Auto Construction Sequence Load Case** tool from the **Define** menu. The **Auto Construction Sequence Load Case** dialog box will be displayed.
- Select the **Case is Active** check box to activate the case and options in this dialog box. The dialog box will be displayed as shown in Figure-30.

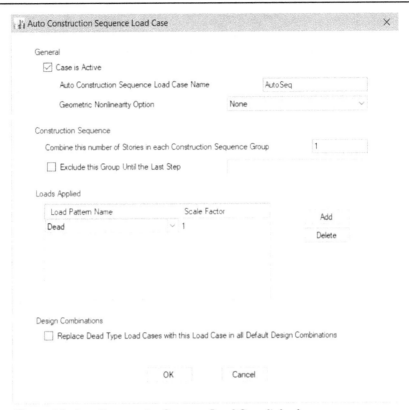

Figure-30. Auto Construction Sequence Load Case dialog box

- Specify the desired name of the load case in the **Auto Construction Sequence Load Case Name** edit box.
- Select the desired option from the **Geometric Nonlinearity Option** drop-down to set the geometric non-linearity effect on all elements in the structure. Select the **None** option so that all equilibrium equations are considered in the undeformed configuration of the structure. Select the **P-Delta** option so that the equilibrium equations take into partial account the deformed configuration of the structure. Tensile forces tend to resist the rotation of elements and stiffen the structure, and compressive forces tend to enhance the rotation of elements and destabilize the structure. This may require a moderate amount of iteration. Select the **P-Delta plus Large Displacements** option so that all equilibrium equations are written in the deformed configuration of the structure. This may require a large amount of iteration.
- Specify the desired value for number of stories to be combined in each construction sequence group in the **Combine this number of Stories in each Construction Sequence Group** edit box. Enter the value for the number of stories in each sequence. This means the structure will start with n stories in the first stage, n more stories will be added in the second stage for a total of 2n stories, and so on.
- Select the **Exclude this Group Until the Last Step** check box to exclude the group selected in the adjacent drop-down until last step analysis is performed.
- Select the desired load pattern from the **Load Pattern Name** column and set the scale factor. Click on the **Add** button to add more loads.

- Select the **Replace Dead Type Load Cases with this Load Case in all Default Design Combinations** check box to set the combination created in this dialog box as default load combination.
- After setting the desired parameters, click on the **OK** button.

CREATING WALKING VIBRATIONS

Walking vibrations are caused when people walk in the building. The procedure to create walking vibration load is given next.

- Click on the **Walking Vibrations** tool from the **Define** menu. The **Walking Vibrations** dialog box will be displayed; refer to Figure-31.

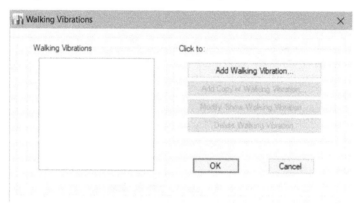

Figure-31. Walking Vibrations dialog box

- Click on the **Add Walking Vibration** button from the dialog box. The **Walking Vibration Data** dialog box will be displayed; refer to Figure-32.

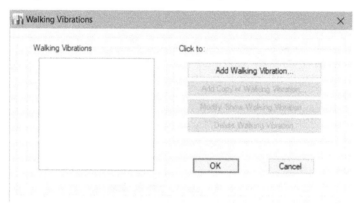

Figure-32. Walking Vibration Data dialog box

- Set the desired name in the **Name** edit box.
- Select the desired story from the **Story** drop-down where you want to apply walking vibrations.
- Set the desired color for displaying walking vibration load case and specify desired notes.
- Select the **Program Default** radio button so that program creates a modal case with the default P-delta using two Ritz modes for each step of the walking vibration case. This option uses the mass as defined in the Default Mass Source so make sure it includes vertical mass. Specify the desired value for Ritz modes per step in **Ritz Modes per Step** edit box.
- Select the **User Specified** radio button to set user defined modal case and damping. Select the desired option from the adjacent drop-down and specify value of modal damping ratio in the **Modal Damping Ratio** edit box.
- Double-click in the desired field of table in **Walking Path** area of the dialog box to edit a walking path point location. Click on the **Add** button from the area to add a new walking path point.
- Specify the average weight of person walking through the building in the **Weight of Person Walking** edit box.
- Set the desired value of peak load factor in the **Peak Load Factor** edit box to define the maximum load exerted due to walking.
- Specify the number of walking steps per second value in the **Walking Frequency (Steps/sec)** edit box.
- Specify the desired value of moving speed in the **Forward Speed** edit box.
- Set the duration in seconds up to which there will be impact of foot steps on the floor in **Duration of Impact** edit box.
- Select the desired radio button from the **Peak Acceleration Threshold (Percentage of Gravity)** area of the dialog box.
- After setting desired parameters, click on the **OK** button. A new walking vibration load will be added in the **Walking Vibrations** list box of **Walking Vibrations** dialog box. After setting desired number of walking vibrations, click on the **OK** button to apply loads.

CHECKING MODEL FOR ANALYSIS

The **Check Model** tool in the **Analyze** menu is used to check if all the parameters have been specified for performing analysis on the model. The procedure to use this tool is given next.

- Click on the **Check Model** tool from the **Analyze** menu. The **Check Model** dialog box will be displayed; refer to Figure-33.
- Select the check boxes from various Checks areas.
- Select the **Trim or Extend Frames and Move Joints to Fix Problems** check box to perform trimming and extending of frames for frame members that overlap, and move joints to fix problem.
- Select the **Joint Story Assignment** check box to assign joints to stories in the building automatically.

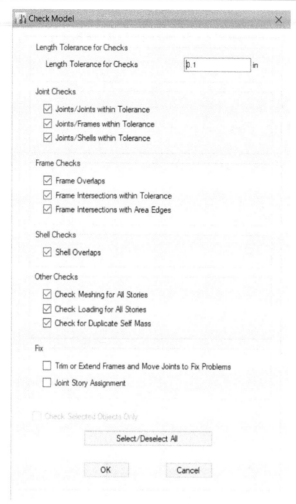

Figure-33. Check Model dialog box

- After setting the desired parameters, click on the **OK** button. The **Warning** box will be displayed with warnings if there is any warning in the model otherwise the **Warning** box will be displayed as shown in Figure-34.

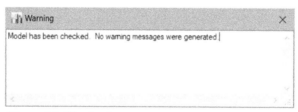

Figure-34. Warning box

SETTING DEGREE OF FREEDOM FOR ANALYSIS

The **Set Active Degrees of Freedom** tool is used to set the degree of freedom up to which the analysis will be performed on the building. The procedure to use this tool is given next.

- Click on the **Set Active Degrees of Freedom** tool from the **Analyze** menu. The **Active Degrees of Freedom** dialog box will be displayed; refer to Figure-35.

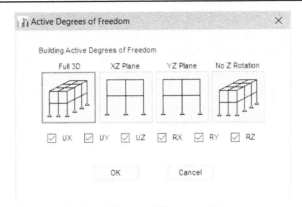

Figure-35. Active Degrees of Freedom dialog box

- Select the **Full 3D** button if you want to run the analysis in 3D.
- Select the **XZ Plane** or **YZ Plane** button to run a 2D analysis in XZ or YZ plane respectively.
- Select the **No Z Rotation** button if you want to perform analysis in 3D environment where model cannot rotate about Z axis but can translate and rotate in all other directions.
- After selecting desired button, click on the **OK** button from the dialog box.

SETTING LOAD CASES TO RUN ANALYSIS

The **Set Load Cases To Run** tool is used to run selected load cases. The procedure to use this tool is given next.

- Click on the **Set Load Cases To Run** tool from the **Analyze** menu. The **Set Load Cases to Run** dialog box will be displayed; refer to Figure-36.

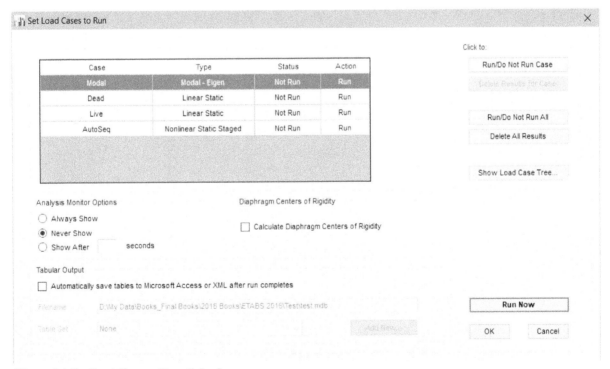

Figure-36. Set Load Cases to Run dialog box

- Select the case that you want to run or do not run from the table. Click on the **Run/Do Not Run Case** button to switch between running and not running the selected load case.
- Click on the **Run/Do Not Run All** button to switch between running and not running all the load cases.
- Select the desired radio button from the **Analysis Monitor Options** area of the dialog box. Select the **Always Show** radio button to display the progress of analysis in the **Analysis** window. Select the **Show After** radio button and specify the time after which the **Analysis** window will be displayed in the adjacent edit box. Select the **Never Show** radio button to do not show the progress of analysis.
- Select the **Calculate Diaphragm Centers of Rigidity** check box to calculate the center of rigidity during the analysis. The original concept of center of rigidity dates back to manual rigidity analysis techniques associated with the lateral analysis of single-story shear wall buildings. Modern computer techniques do not require the explicit evaluation of the center of rigidity. However, the center of rigidity still needs to be evaluated because some building codes refer to it as a reference point to define design eccentricity requirements in multistory buildings.
- If you want to save data specified in this table in tabular form in a Microsoft Access or XML file then select the **Automatically save tables to Microsoft Access or XML after run completes** check box. Specify the desired file name and location of file in the **Filename** edit box.
- Click on the **OK** button to set the load cases.

SETTING SAPFIRE SOLVER OPTIONS

The **Advanced SAPFire Options** tool is used to set the solver for analysis. The procedure to use this tool is given next.

- Click on the **Advanced SAPFire Options** tool from the **Analyze** menu. The **Advanced SapFire Options** dialog box will be displayed; refer to Figure-37.

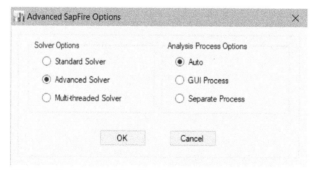

Figure-37. Advanced SapFire Options dialog box

- Select the **Standard Solver** radio button if you want to use only single core and check for model stability. This solver is used for smaller problems.
- Select the **Advanced Solver** radio button if you want to run analysis under default settings for medium to large problems. This option uses all cores of your CPUs.

- Select the **Multi-threaded Solver** radio button to solve medium to large problems. This option utilizes all CPU cores and full RAM during the analysis.
- Select the desired options from the **Analysis Process Options** area. Description of these options is given next.

GUI process
- Best for small problems
- Analysis runs within the software, such as with SAFE.exe
- Benefit: less disk operations (I/O) are performed
- Drawback: the software itself consumes memory, leaving less available for analysis, which slows operations and prevents the ability to run larger models

Separate process
- Best for medium to large problems
- The analysis model is written to the disk and read by CSI.SAPFire. Driver.exe, then analysis is run within CSI.SAPFire.Driver.exe
- Benefit: the analysis engine has access to more memory, therefore larger problems can be solved and analysis runs faster
- Drawback: time is lost to write and read the analysis model

Auto
- Default setting
- The necessary memory is estimated, then compared to the physical ram available. If enough RAM is available, the analysis runs in GUI process. If not, it is shelled out to CSI.SAPFire.Driver.exe.

- After setting the desired parameters in the dialog box and click on the **OK** button.

AUTOMATIC MESH SETTINGS FOR FLOORS

The **Automatic Mesh Settings for Floors** tool is used to set mesh size for floors in the building. The procedure to use this tool is given next.

- Click on the **Automatic Mesh Settings for Floors** tool from the **Analyze** menu. The **Automatic Mesh Options** dialog box will be displayed; refer to Figure-38.
- Set the desired mesh size in the **Approximate Maximum Mesh Size** edit box in feet.
- Set the desired parameters and click on the **OK** button. The specified size of mesh will be applied.

Figure-38. Automatic Mesh Options (for Floors) dialog box

SETTING MESH SIZE FOR WALLS

The **Automatic Rectangular Mesh Settings for Walls** tool is used to specify the size of mesh elements for walls. The procedure to use this tool is given next.

- Click on the **Automatic Rectangular Mesh Settings for Walls** tool from the **Analyze** menu. The **Automatic Rectangular Mesh Options (for Walls)** dialog box will be displayed; refer to Figure-39.

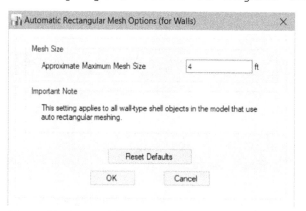

Figure-39. Automatic Rectangular Mesh Options (for Walls) dialog box

- Specify the desired mesh size in the edit box and click on the **OK** button.

SETTING HINGES FOR ANALYSIS MODEL

The **Analysis Model for Nonlinear Hinges** tool is used to define how hinges will work in the analysis model. The procedure to use this tool is given next.

- Click on the **Analysis Model for Nonlinear Hinges** tool from the **Analyze** menu. The **Analysis Model for Nonlinear Hinges** dialog box will be displayed; refer to Figure-40.

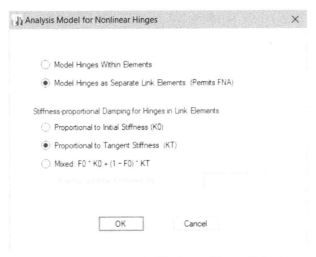

Figure-40. Analysis Model for Nonlinear Hinges dialog box

- Select the **Model Hinges Within Elements** radio button to make hinges integral part of frame and wall objects.
- Select the **Model Hinges as Separate Link Elements (Permits FNA)** radio button to model hinges with frames and walls using an additional link element. These hinges may be used in all nonlinear analyses, including the often faster and more stable Fast Nonlinear Analysis (FNA). On selecting this radio button, the radio buttons in the **Stiffness-proportional Damping for Hinges in Link Elements** area will become active.
- Select the desired radio button to define damping for hinges.
- After setting the desired parameters, click on the **OK** button.

CRACKING ANALYSIS OPTIONS

The **Cracking Analysis Options** tool is used to set options for performing cracking analysis. The procedure to use this tool is given next.

- Click on the **Cracking Analysis Options** tool from the **Analyze** menu. The **Reinforcement Options For Cracking Analysis** dialog box will be displayed; refer to Figure-41.
- Select the desired option from the **Reinforcement Source** area to define source of reinforcement bars.
- Specify the desired minimum reinforcing ratio for tension and compression in the respective edit boxes. Based on the specified value, the cracking analysis will be performed.

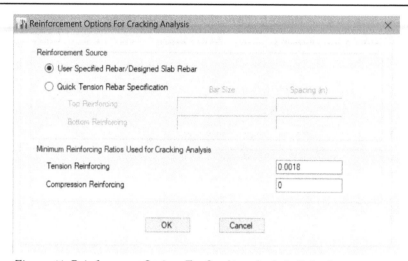

Figure-41. Reinforcement Options For Cracking Analysis dialog box

RUNNING THE ANALYSIS

The **Run Analysis** tool in **Analyze** menu is used to perform analysis. The procedure to perform analysis is given next.

- After setting all the parameters, click on the **Run Analysis** tool from the **Analyze** menu. The analysis will be performed and result will be displayed in selected viewport; refer to Figure-42.

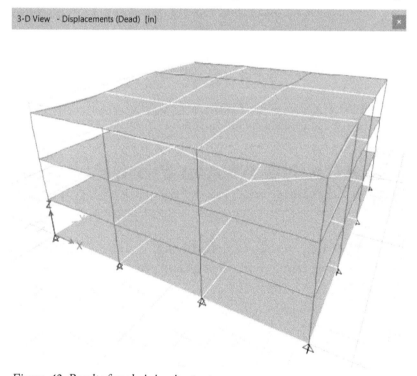

Figure-42. Result of analysis in viewport

APPLYING MODEL ALIVE

The **Model Alive** tool in the **Analyze** menu is a toggle button used to perform analysis automatically when the model is changed. The procedure is given next.

• Click on the **Model Alive** tool from the **Analyze** menu to toggle it on. The analysis mode will become active. Change the model as required and you will see the results accordingly.

MODIFYING UNDEFORMED GEOMETRY

The **Modify Undeformed Geometry** tool is active once you have performed the analysis. This tool is used to modify the model in its undeformed shape. The procedure to use this tool is given next.

• Click on the **Modify Undeformed Geometry** tool from the **Analyze** menu. The **Modify Undeformed Geometry** dialog box will be displayed; refer to Figure-43.

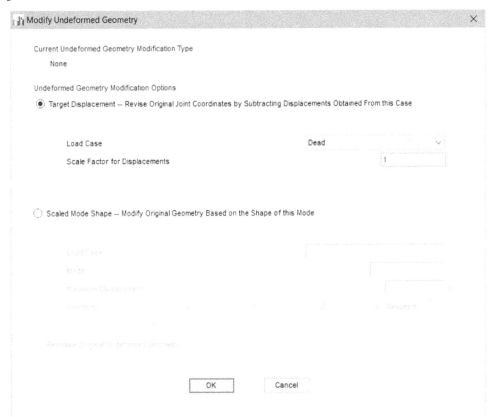

Figure-43. Modify Undeformed Geometry dialog box

• Selecting the **Target Displacement** radio button modifies the original geometry of the structure in order to achieve a desired deformed shape. When this option is selected, use the **Load Case** drop-down list to select the load case for which the geometry is to be modified.

• Selecting the **Scaled Mode Shape** radio button modifies the geometry of the structure based on a specified modal or buckling load case mode shape. This option is useful for modifying the input geometry to consider things such as imperfections. When this option is selected, use the **Load Case** drop-down list to select the desired load case and the **Mode** drop-down list to select the associated mode. The maximum Displacement value represents the maximum change to be applied to the geometry, in the specified direction. The mode shape is scaled to the specified maximum displacement in the specified direction and used to modify the geometry of the structure. In order to modify the geometry based on another mode shape or to a different maximum displacement, it is necessary to first reinstate the original geometry.

• After setting the desired parameters, click on the **OK** button.

CHECKING ANALYSIS LOG

The **Last Analysis Run Log** tool is used to check the details of analysis recently performed. To check the log, click on the **Last Analysis Run Log** tool from the **Analyze** menu. The **Analysis Complete** window will be displayed; refer to Figure-44.

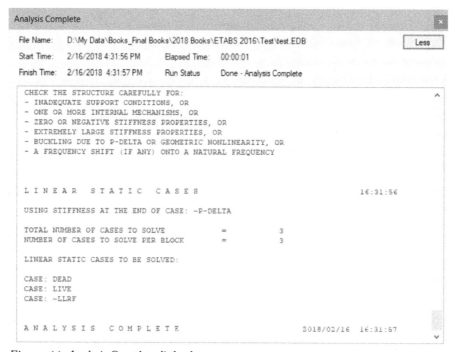

Figure-44. Analysis Complete dialog box

UNLOCKING MODEL

Once the analysis is performed, the model gets locked for any changes. To perform changes, click on the **Unlock Model** button from the **Analyze** menu.

DISPLAY OPTIONS

The tools in the **Display** menu are used to display various properties of the model.

Undeformed Shape

The **Undeformed Shape** tool is used to view the model in undeformed shape in selected viewport. To use this tool, select the desired viewport and click on the **Undeformed Shape** tool from the **Analyze** menu.

Load Assigns

The options in the **Load Assigns** cascading menu of **Display** menu are used to display various load assigns on the model in selected viewport; refer to Figure-45.

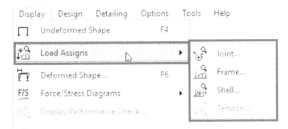

Figure-45. Load Assigns cascading menu

Select the desired option from the cascading menu to display respective load assign in the viewport.

Deformed Shape

The **Deformed Shape** tool in the **Display** menu is used to display model in selected viewport as deformed. To use this tool, select the viewport and click on the **Deformed Shape** tool. The model will be displayed in deformed shape.

Force/Stress Diagrams

The tools in the **Force/Stress Diagrams** cascading menu are used to display forces, reactions, and stress on the model; refer to Figure-46. The tools in this menu are discussed next.

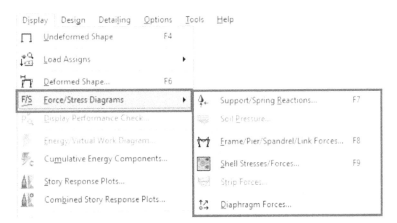

Figure-46. Force Stress Diagrams cascading menu

Support/Spring Reactions

The **Support/Spring Reactions** tool in the **Force/Stress Diagrams** cascading menu is used to display reaction forces of support/spring in the model. The procedure to use this tool is given next.

- Select the desired viewport in which you want to display reaction forces. Click on the **Support/Spring Reactions** tool in **Force/Stress Diagrams** cascading menu of the **Display** menu. The **Reactions** dialog box will be displayed; refer to Figure-47.

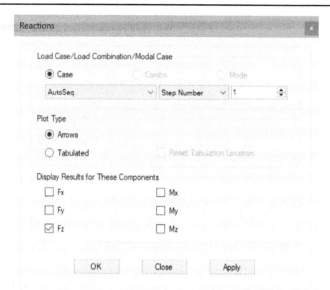

Figure-47. Reactions dialog box

- Select the **Case** radio button to check reaction force of selected load case. Select the **Combo** radio button to check the reaction force of selected load combination. Select the **Mode** radio button to check the reaction force of selected modal case. After selecting the desired radio button, set the options in drop-downs of **Load Case/Load Combination/Modal Case** area.

- Select the desired radio button from the **Plot Type** area to define how reaction forces will be displayed in the viewport. Select the **Arrows** radio button to display reaction forces in the form of arrows. Select the **Tabulated** radio button to display reaction forces in the form of tabular data. Select the **Reset Tabulation Location** check box to reset the locations earlier set in the viewport.

- Select the check boxes for the components that you want to display in the viewport.

- After setting desired parameters, click on the **OK** button to display reaction forces.

Displaying Soil Pressure

The **Soil Pressure** tool in the **Force/Stress Diagrams** cascading menu is used to display soil pressure on the model. The procedure is given next.

- Click on the **Soil Pressure** tool from the **Force/Stress Diagrams** cascading menu of the **Display** menu. The **Display Soil Pressure** dialog box will be displayed; refer to Figure-48.

- Select the desired radio button from the **Load Case/Load Combination/Modal Case** area and set the options in the drop-downs below it.

- Select the desired option in **Contour Option** drop-down to define how soil pressure will be displayed in the model. Select the **Display on Undeformed Shape** option if you want to display soil pressure on undeformed model. Select the **Display on Deformed Shape** option if you want to display soil pressure in deformed shape. Select the **Display in Extruded Form** option if you want to display extruded contours of soil pressure.

Figure-48. Display Soil Pressure dialog box

- Set the desired parameters and click on the **OK** button. The soil pressure plot will be displayed; refer to Figure-49.

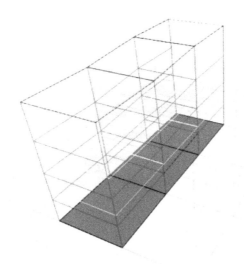

Figure-49. Soil pressure plot

Frame/Pier/Spandrel/Link Forces

The **Frame/Pier/Spandrel/Link Forces** tool is used to display component of selected force or moment in the selected viewport. The procedure is given next.

- Click on the **Frame/Pier/Spandrel/Link Forces** tool from the **Force/Stress Diagrams** cascading menu of the **Display** menu. The **Member Force Diagram for Frames/Piers/Spandrels/Links** dialog box will be displayed; refer to Figure-50.

Figure-50. Member Force Diagram for Frames/Piers/Span-drels/Links dialog box

- Select the desired option from the **Component** area to display respective force/moment component in the model.
- Select the objects to be included in the diagram to display forces/ moments from the **Include** area.
- Set the other parameters as discussed earlier and click on the **OK** button. The forces/moments will be displayed in the selected viewport.

Shell Stresses/Forces

The **Shell Stresses/Forces** tool is used to display the stresses/forces exerted on the shell members. The procedure to use this tool is given next.

- Click on the **Shell Stresses/Forces** tool from the **Force/Stress Diagrams** cascading menu of the **Display** menu. The **Shell Forces/Stresses** dialog box will be displayed; refer to Figure-51.
- Set the parameters as discussed earlier and click on the **OK** button.

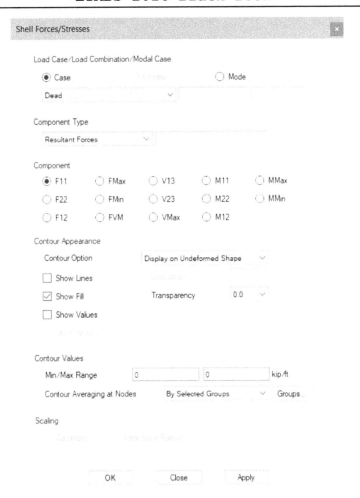

Figure-51. Shell Forces/Stresses dialog box

Similarly, you can use the **Strip Forces** and **Diaphragm Forces** tools.

Display Performance Check

The **Display Performance Check** tool is used to displays demand/capacity (D/C) ratios for hinges after a nonlinear time history analysis has been run.

Energy/Virtual Work Diagram

The **Energy/Virtual Work Diagram** tool is used to display energy diagrams that can be used as an aid to determine which elements should be stiffened to achieve the most efficient control over the lateral displacements of a structure. The procedure to use this tool is given next.

• Click on the **Energy/Virtual Work Diagram** tool from the **Display** menu. The **Energy/Virtual Work Diagram** dialog box will be displayed; refer to Figure-52.

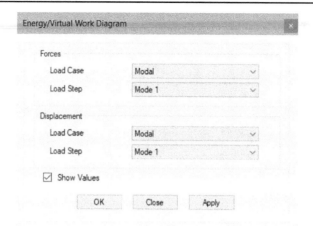

Figure-52. Energy/Virtual Work Diagram dialog box

- Select the desired load cases and load steps from the **Forces** and **Displacement** areas.
- Click on the **OK** button. The viewport will be displayed with forces and displacements annotated.

Cumulative Energy Components

The **Cumulative Energy Components** tool plots all energy components in a cumulative manner after a modal time history analysis has been run. Click on the **Cumulative Energy Components** tool from the **Display** menu. The **Cumulative Energy Components** dialog box will be displayed; refer to Figure-53.

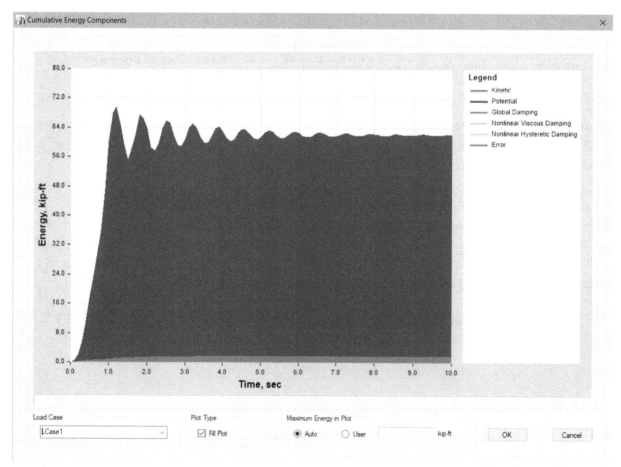

Figure-53. Cumulative Energy Components dialog box

Select the desired load case from the **Load Case** drop-down. Set the other options as required and click on the **OK** button.

Story Response Plots

The **Story Response Plots** tool is used to display force and displacement responses for specified stories as a new tab in a display window. The procedure to use this tool is given next.

- Click on the **Story Response Plots** tool from the **Display** menu. The **Story Response** window will be displayed; refer to Figure-54.

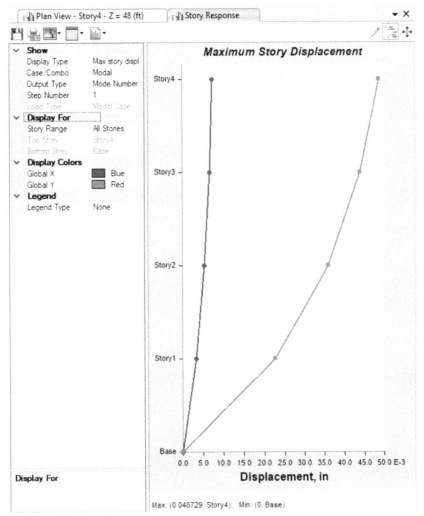

Figure-54. Story Response window

- Set the desired parameters in the table at the left of the graph.

Plot Functions

The **Plot Functions** tool in the **Display** menu is used to plot time history graph of the building. The procedure to use this tool is given next.

- After performing the time history analysis, click on the **Plot Functions** tool from the **Display** menu. The **Time History Plot** window will be displayed; refer to Figure-55.

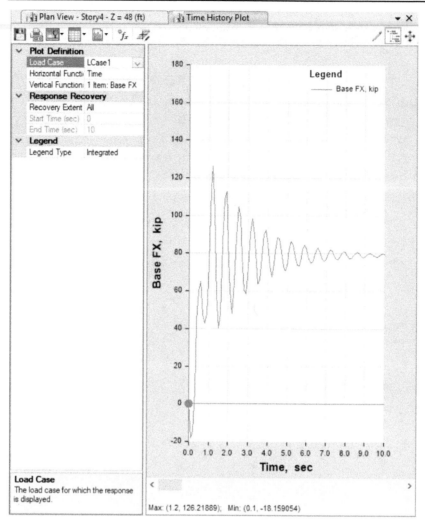

Figure-55. Time History Plot window

- Set the desired parameters in the table at the left of the window and check the plot.
- Close the window once you have checked the plot.

You can create the other plots in the same way.

Once you have performed the analysis then its time to perform design study so that size and parameters can be updated as per the analysis results.

Chapter 6

Running Design Study

The major topics covered in this chapter are:

- *Introduction*
- *Steel Frame Design*
- *Concrete Frame Design*
- *Composite Beam/Column Design*
- *Overwrite Frame Design Procedure*
- *Shear Wall Design and Concrete Slab Design*
- *Live Load Reduction Factors*
- *Setting Lateral Displacement Targets*
- *Setting Time Period Targets*

INTRODUCTION

In previous chapters, you have learned to draw frame members and create properties of different frame members. In this chapter, you learn to set preferences for the project and assign properties to various frame members.

STEEL FRAME DESIGN

The parameters related to steel frame design are available in the **Steel Frame Design** cascading menu of the **Design** menu; refer to Figure-1. Various options in this menu are discussed next.

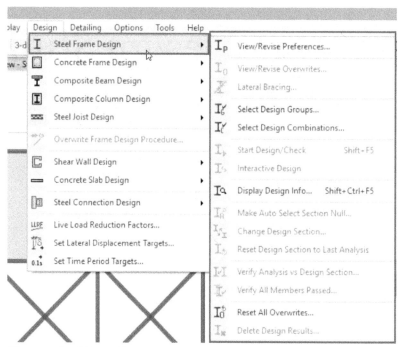

Figure-1. Steel Frame Design cascading menu

Steel Frame Preferences

Steel Frame Preferences are the set of rules to be abide by all steel frame members in the project. The procedure to set these rules is given next.

• Click on the **View/Revise Preferences** tool from the **Design** menu. The **Steel Frame Design Preferences** dialog box will be displayed; refer to Figure-2.
• Select the desired design code from the **Design Code** field in the table. There are various standards as per the construction industry.
• Select the desired option from the **Multi-Response Case Design** field. The **Step-by-Step-All** is used when analysis is to be performed step by step for first Time History, then Multi-step static and at the end Nonlinear static. Select the **Envelopes-All** option if you want to run the analysis all along and compile a final result. Select the **Last Step** option if you want to check the considers last values for Time History, Multi-step static and Nonlinear static analysis. You can use the other options in the same way.

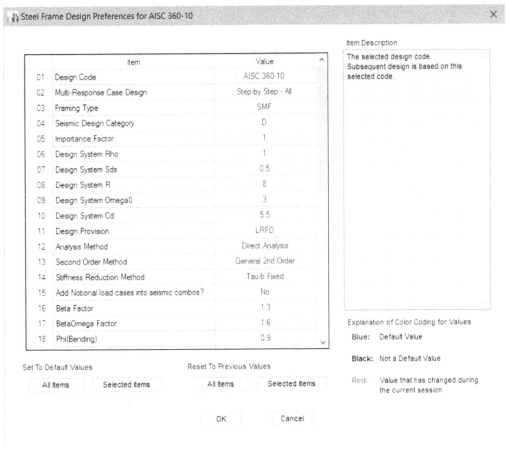

Figure-2. Steel Frame Design Preferences for AISC 360-10 dialog box

- Select the desired option from the **Framing Type** drop-down. Various parameters in the drop-down and their respective references are given in the next table.

Framing Type	References
SMF (Special Moment Frame)	AISC SEISMIC 9
IMF (Intermediate Moment Frame)	AISC SEISMIC 10
OMF (Ordinary Moment Frame)	AISC SEISMIC 11
STMF (Special Truss Moment Frame)	AISC SEISMIC 12
SCBF (Special Concentrically Braced Frame)	AISC SEISMIC 13
OCBF (Ordinary Concentrically Braced Frame)	AISC SEISMIC 14
EBF (Eccentrically Braced Frame)	AISC SEISMIC 15
BRBF (Buckling Restrained Braced Frame)	AISC SEISMIC 16
SPSW (Special Plate Shear Wall)	AISC SEISMIC 17

- Select the desired option from the **Seismic Design Category** drop-down. These options are related to seismic design.
- Set the desired value of importance factor in the Importance Factor edit box.
- Similarly, you can set the desired value of redundancy factor (Design System Rho), Design Spectral Acceleration Parameter (Design System Sds), Response Modification Factor (Design System R),

---

System Overstrength Factor (Design System Omega0), and Deflection Amplification Factor (Design System Cd) in their respective edit boxes.

- Select the desired option from the **Design Provision** drop-down. The first method is the Allowable Strength Design (ASD) method. The second is the Load and Resistance Factor Design (LRFD) method.
- Select the desired analysis method from the **Analysis Method** drop-down. There are three options in this drop-down; **Direct Analysis**, **Effective Length**, and **Limited 1st Order**. The **Effective Length** method is a traditional analysis method used to solve small second order effects of load on the building. In the AISC 360-05/IBC 2006 code, the effective length method is allowed provided the member demands are determined using a second-order analysis (either explicit or by amplified first-order analysis) and notional loads are included in all gravity load combinations. K-factors must be calculated to account for buckling (except for braced frames, or where Δ2 / Δ1 < 1.0, K = 1.0). The **Direct Analysis Method** is expected to more accurately determine the internal forces of the structure, provided care is used in the selection of the appropriate methods used to determine the second-order effects, notional load effects and appropriate stiffness reduction factors as defined in AISC 2.2, App. 7.3(3). Additionally, the Direct Analysis Method does not use an effective length factor other than k = 1.0. The rational behind the use of k = 1.0 is that proper consideration of the second-order effects (P-Δ and P-δ), geometric imperfections (using notional loads) and inelastic effects (applying stiffness reductions) better accounts for the stability effects of a structure than the Effective Length method. Use the Limited 1st Order method when the analysis is for relatively simple problems. The **Limited First Order Analysis** does not include the secondary P-Δ and P-δ effects. This method has very limited applicability and might be appropriate only when the axial forces in the columns are very small compared to their Euler buckling capacities.
- Select the desired method from the **Second Order Method** field in the table. These methods are only useful when you have selected **Direct Analysis** or **Effective Length** option from the **Analysis Method** field.
- Select the desired option from the **Stiffness Reduction Method** drop-down in the table. There are three options in this drop-down; **Tau-b Variable**, **Tau-b Fixed**, and **No Modification**. Select the **Tau-b Variable** or **Tau-b Fixed** option if you want the apply reduction factor.
- Select the **Yes** option from the **Add Notional load cases into seismic combos** drop-down if you want to add the notional loads to seismic load desired from standards.
- Similarly, set the other parameters in the table and click on the **OK** button.

View/Revise Overwrite for Steel Frame Sections

The **View/Revise Overwrite** tool in the **Steel Frame Design** cascading menu is used to overwrite the default preferences for selected steel frame section. The procedure to use this tool is given next.

- Select the desired steel section from the model and click on the **View/Revise Overwrite** tool from the **Steel Frame Design** cascading menu of the **Design** menu. The **Steel Frame Design Overwrites** dialog box will be displayed; refer to Figure-3.

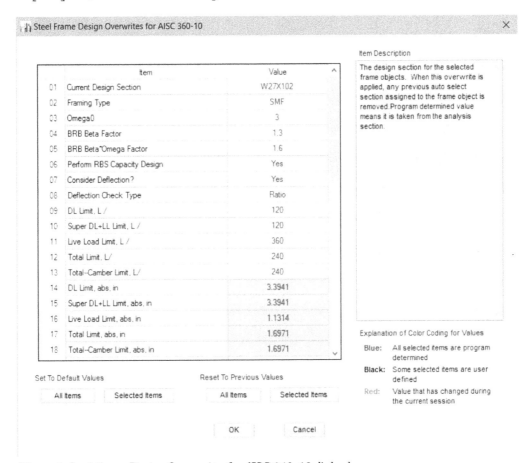

Figure-3. Steel Frame Design Overwrites for AISC 360-10 dialog box

- Set the desired parameters in the table and click on the **OK** button.

Setting Lateral Brace Parameters

Lateral bracing is the term we use to refer to any pieces on a bridge that help keep the sides (trusses) from twisting. It also helps keep the top chords of the bridge from bending or deforming in or out. In Figure-4, the lateral bracing is marked red. The procedure to set lateral bracing is given next.

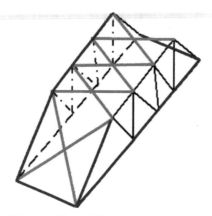

Figure-4. Lateral bracing

- Click on the **Lateral Bracing** tool from the **Steel Frame Design** cascading menu of the **Design** menu. The **Lateral Bracing** dialog box will be displayed; refer to Figure-5.

Figure-5. Lateral Bracing dialog box

- Select the **Program Determined** radio button if you want the system to automatically set parameters for lateral bracing.
- To manually set the parameters, click on the **User Specified** radio button. The buttons below it will become active.
- Click on the **Specify Point Bracing** button if you want to create bracing at desired points. The **Point Braces** dialog box will be displayed; refer to Figure-6.

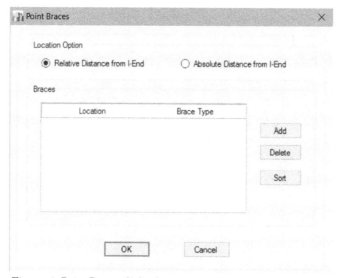

Figure-6. Point Braces dialog box

- Select the **Relative Distance from I-End** radio button if you want to specify the distance in ratio of the total length. Select the **Absolute Distance from I-End** radio button if you want to specify the absolute value of distance from the end point of selected beam.
- Click on the **Add** button from the dialog box. A new location of brace will be added in the table.
- Specify the desired value of location in the **Location** field and set the brace type.
- Click on the **OK** button to apply braces.
- Similarly, you can create uniform bracing by using the **Specify Uniform Bracing** button.
- After creating the braces, click on the **OK** button from the dialog box.

Selecting Design Group

The **Select Design Group** tool is used to select a design group earlier created by using the **Group Definitions** tool in the **Define** menu for select components of structure. The procedure is given next.

- Click on the **Select Design Groups** tool from the **Steel Frame Design** cascading menu of the **Design** menu. The **Steel Frame** dialog box will be displayed as shown in Figure-7.

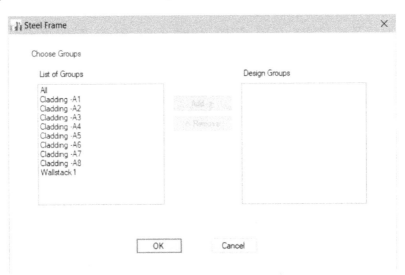

Figure-7. Steel Frame dialog box

- Double-click on the design groups that you want to be included in the current project and click on the **OK** button. The selected groups will be available for assignment to various structural components.

Selecting Design Combinations

The **Select Design Combinations** tool is used to select the design load combinations applicable in current project. The procedure to use this tool is given next.

- Click on the **Select Design Combinations** tool from the **Steel Frame Design** cascading menu of the **Design** menu. The **Design Load Combinations Selection** dialog box will be displayed; refer to Figure-8.

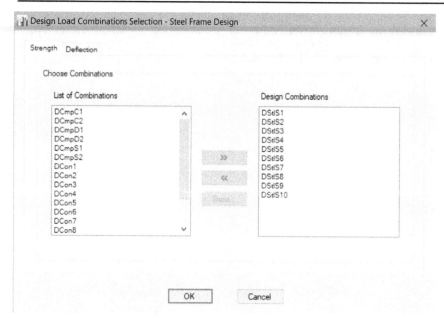

Figure-8. Design Load Combinations Selection dialog box

- There are two load groups in the dialog box viz. **Strength** and **Deflection**. Select the **Strength** tab in the dialog box if you want to use strength based design load combinations in your project. Select the **Deflection** tab if you want use the deflection based design load combinations in your project.
- Double-click on the design load combination from the **List of Combinations** list box in the dialog box to add the selected combination in the project.
- Click on the **OK** button from the dialog box to apply the changes.

Performing Design Check

The **Start Design/Check** tool is used to perform design checks on the structure based on specified load combinations and selected design groups. This tool is available when you have performed analysis by using the **Run Analysis** tool in the **Analyze** menu. You need to check the topics of Load assignment and performing analysis in next chapters before using this tool. The procedure to use this tool is given next.

- Click on the **Start Design/Check** tool from the **Steel Frame Design** cascading menu in the **Design** menu. The **Steel Design/Check** information box will be displayed while the calculations are being performed. Once the analysis is done the results of design check will be displayed in the selected viewport; refer to Figure-9.

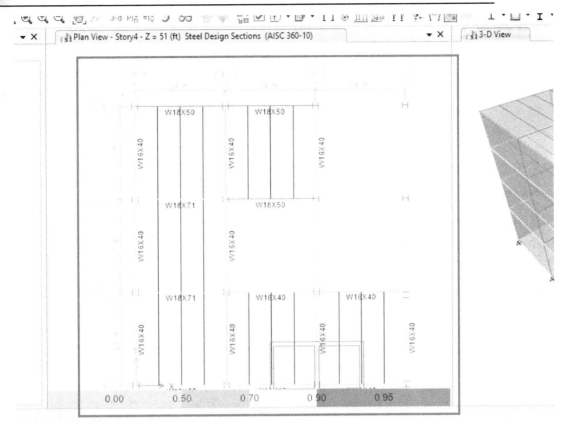

Figure-9. Results of design check

In the design results, each member of structure is displayed with its dimensional name and colors are assigned to them based on their load conditions. Like the structural members in red color are showing high deflection as compared to structural members in cyan color.

Stress Information of Structural members

Once you have performed the design check, it is required to check the details of stress induced on high deflecting components. To do so, follow the steps given next.

• To check this information, click on the desired structural member from the design check result and right-click on it. The **Steel Stress Check Information** dialog box will be displayed; refer to Figure-10.

The details of each column in the table are as follows:

COMBO ID: The name of the design load combination considered.

STATION LOC: The location of the station considered, measured from the i-end of the frame element.

RATIO: The total PMM stress ratio for the element. When stress ratios are reported for this item, they are followed by (T) or (C). The (T) indicates that the axial component of the stress ratio is tension. The (C) indicates that the axial component of the stress ratio is compression. Note that typically the interaction formulas are different depending on whether the axial stress is tension or compression.

AXL: The axial component of the PMM stress ratio.

B-MAJ: The bending component of the PMM stress ratio for bending about the major axis.

B-MIN: The bending component of the PMM stress ratio for bending about the minor axis.

MAJ SHR RATIO: The shear stress ratio for shear acting in the major direction of the frame element.

MIN SHR RATIO: The shear stress ratio for shear acting in the minor direction of the frame element.

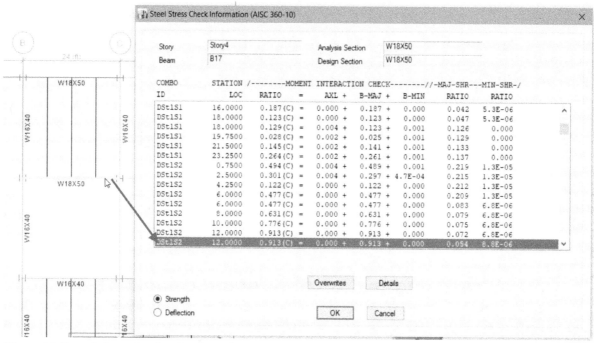

Figure-10. Stress information for selected member

- To check further details of stress on selected structure member, select it from the table and click on the **Details** button. Complete summary of the selected member will be displayed; refer to Figure-11.
- Click on the **Envelope** tab to check the strength under envelope type load combinations.
- Close the dialog box and click on the **OK** button from the **Steel Stress Check Information** dialog box to exit.

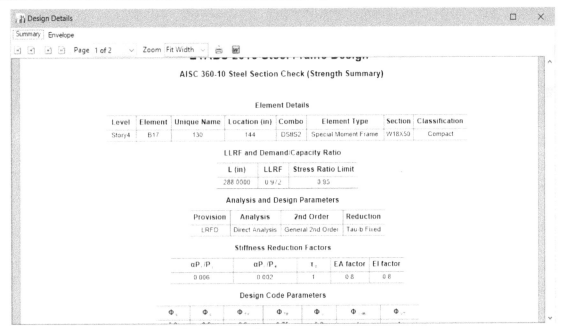

Figure-11. Strength summary of selected structural member

Interactive Design View

While designing the structure, you can check the interactive view of the frame members in shades according to their deformation under specified standard load conditions. To activate this tool, select the desired view and click on the **Interactive Design** tool from the **Steel Frame Design** cascading menu in the **Design** menu. The interactive design view will be displayed; refer to Figure-12.

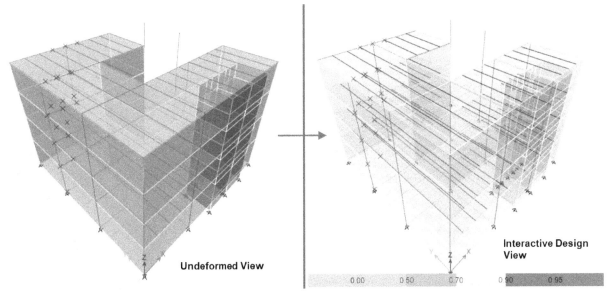

Figure-12. Interactive design view

Display Design Info

The **Display Design Info** tool is used to check different design parameters of the selected structure member. The procedure to use this tool is given next.

- Select the desired view in which you want to display the design parameters and click on the **Display Design Info** tool from the **Steel Frame Design** cascading menu of the **Design** menu. The **Display Steel Frame Design Results** dialog box will be displayed; refer to Figure-13.

Figure-13. Display Steel Frame Design Results dialog box

- Select the desired radio button to display design output parameters or design input parameters.
- Click in the drop-down and select the parameter you want to display in the view.
- Click on the **OK** button to apply the modifications.

Removing Auto Select Sections

The **Make Auto Select Section Null** tool to clear the list of auto select sections after performing analysis. The procedure to use this tool is given next.

- Click on the **Make Auto Select Section Null** tool from the **Steel Frame Design** cascading menu of the **Design** menu. The **ETABS** information box will be displayed; refer to Figure-14.

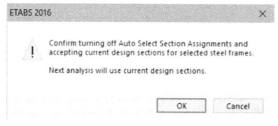

Figure-14. ETABS information box

- Click on the **OK** button from the box to disable auto selection of sections and assign the present section sizes to various members of structure.

Changing Steel Frame Sections

The **Change Design Section** tool is used to change the size of selected sections of the frame. This tool is active only when a frame member is selected. The procedure to use this tool is given next.

- Select the desired frame member and click on the **Change Design Section** tool from the **Steel Frame Design** cascading menu of the **Design** menu. The **Select Sections** dialog box will be displayed; refer to Figure-15.

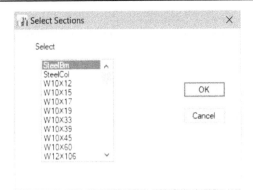

Figure-15. Select Sections dialog box

- Select the desired section and click on the **OK** button from the dialog box to apply the changes.

Resetting the Design Sections to Previous Analysis Values

The **Reset Design Section to Last Analysis** tool is used to reset the sections of frame to previous values generated by last analysis. To do so, click on the **Reset Design Section to Last Analysis** tool from the Steel Frame Design cascading menu of the Design menu. The values will be reset automatically.

Verifying Analysis V/s Design Section

The **Verify Analysis vs Design Section** tool is used to check whether the assigned section values are same as generated by analysis. If there is any section different from generated after analysis then an information box will be displayed. The procedure to use this tool is given next.

- Click on the **Verify Analysis vs Design Section** tool from the **Steel Frame Design** cascading menu of the **Design** menu. If there is no difference between the values of section in model and the values generated by analysis then the information box will be displayed as shown in Figure-16. If there is a difference between the values then the information box will be displayed as shown in Figure-17. Click on the **Yes** button from the dialog box to select all the differing sections.

Figure-16. Information box for matching steel sections   *Figure-17. Information box with differing steel sections*

- Now, you can perform the required operations on the selected frame members.

Verifying All Members according to Analysis Parameters

The **Verify All Members Pass** tool is used to check if all the members of frame have passed the analysis or not. The procedure to use this tool is given next.

* Click on the **Verify All Members Pass** tool from the **Steel Frame Design** cascading menu in the **Design** menu. The information box will be displayed as shown in Figure-18 if there is any member on which the analysis has not been performed. Click on the **Yes** button to select the left members. Perform the analysis again to set parameters as per the analysis. If all the members have passed the analysis then the related information box will be displayed. Click on the **OK** button to exit.

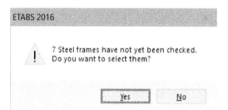

Figure-18. Information box for frame
members not checked

Resetting All Overwrites

The **Reset All Overwrites** tool is used to reset all the parameters of steel frame to design default values. To reset overwrites, click on the **Reset All Overwrites** tool from the **Steel Frame Design** cascading menu of the **Design** menu. The dialog box for resetting parameters will be displayed as shown in Figure-19. Click on the **OK** button from the dialog box to reset steel frame members.

Figure-19. Dialog box for resetting steel frame
parameters

Deleting Design Results

The **Delete Design Results** tool is used to delete the results of previous design check performed.

CONCRETE FRAME DESIGN

The tools in the **Concrete Frame Design** cascading menu are used to design concrete frames based on analysis and design check; refer to Figure-20. The tools in this cascading menu work in the same way as discussed for Steel Frame Design. Note that the tools in this menu will be active only when you can assigned concrete beams/columns to the frame and you have performed the analysis.

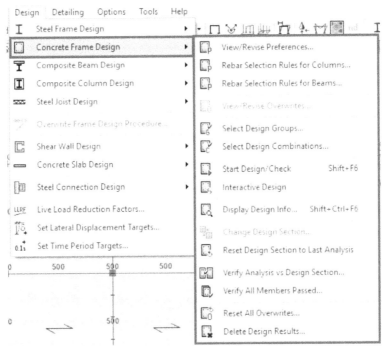

Figure-20. Concrete Frame Design cascading menu

COMPOSITE BEAM DESIGN, COMPOSITE COLUMN DESIGN, AND STEEL JOIST DESIGN

The tools in the **Composite Beam Design**, **Composite Column Design**, and **Steel Joist Design** cascading menus are same as discussed for Steel Frame Design.

OVERWRITE FRAME DESIGN PROCEDURE

The **Overwrite Frame Design Procedure** tool is used to overwrite the design postprocessor for selected frames. Using this tool, you can set composite design postprocessor for selected steel frame members or you can set steel postprocessor for selected composite beam. The procedure to use this tool is given next.

• Select the frame member for which you want to change the design postprocessor and click on the **Overwrite Frame Design Procedure** tool from the **Design** menu. The **Overwrite Frame Design Procedure** dialog box will be displayed; refer to Figure-21.

Figure-21. Overwrite Frame Design
Procedure dialog box

• Select the desired radio button the dialog box and click on the **OK** button. The changes will be applied to selected members.

Note that there is a limitation to overwriting frame design procedures and depending on the frame section assignment made to the frame object, several of the options may not be available. The conditions for this tool are given next.

1. A concrete frame element can be switched between the Concrete Frame Design and the No Design procedures. Assign a concrete frame element the No Design (Null) design procedure if you do not want it designed by the Concrete Frame Design postprocessor.

2. A steel frame element can be switched between the Steel Frame Design, Composite Beam Design (if it qualifies), and the No Design design procedures. In this form a steel frame element qualifies for the Composite Beam Design procedure if it meets all of the following criteria.
    • The frame type is Beam, that is, the frame object is horizontal.
    • The frame element is oriented with its positive local 2-axis in the same direction as the positive global Z-axis (vertical upward).
    • The frame element has I-section or channel section properties.
    • Assign a steel frame element the No Design procedure if you do not want it designed by either the Steel Frame Design or the Composite Beam Design postprocessor.

3. A composite beam element can be switched between the Steel Frame Design and the No Design (see above).

4. A composite column element can be switched to No Design.

5. A steel joist element can be switched to No Design. Only beams that are assigned joist sections can be designed as steel joists. The joist sections can be selected from the built-in program joist section database, or they can be user defined. The user-defined sections can be specified using the Define menu > Frame Sections command and the Steel Joist option.

6. No Design option. This option is available for all frame objects. It means that the frame object is not designed by any design

postprocessor. These frame objects are given the Null design procedure.

7. Default option. This option is available for all frame objects. It means that the frame object is to be given the default design procedure (postprocessor).

SHEAR WALL DESIGN AND CONCRETE SLAB DESIGN

The tools in the **Shear Wall Design** cascading menu are used to design load bearing walls; refer to Figure-22. The tools in the **Concrete Slab Design** cascading menu are used to design concrete slabs; refer to Figure-23. The tools in these cascading menus work in the same way as the tools in **Steel Frame Design** cascading menu.

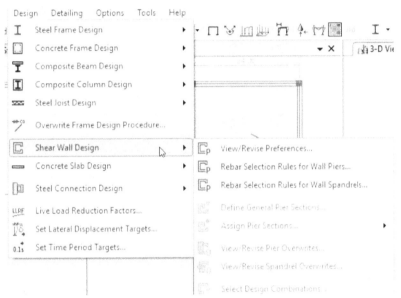

Figure-22. Shear Wall Design cascading menu

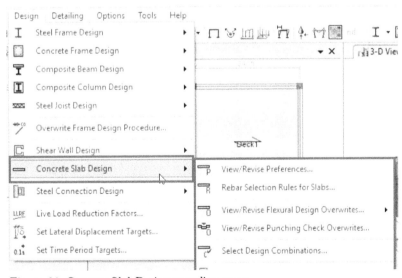

Figure-23. Concrete Slab Design cascading menu

LIVE LOAD REDUCTION FACTORS

The **Live Load Reduction Factors** tool is used to set preferences for reduction in live load during the course of analysis. Note that to use this tool, you must have specified a live reducible load in load patterns (the process to assign reducible live load has been dicussed in previous chapter). The procedure to use this tool is given next.

- Click on the **Live Load Reduction Factors** tool from the **Design** menu. The **Live Load Reduction Factor** dialog box will be displayed; refer to Figure-24.

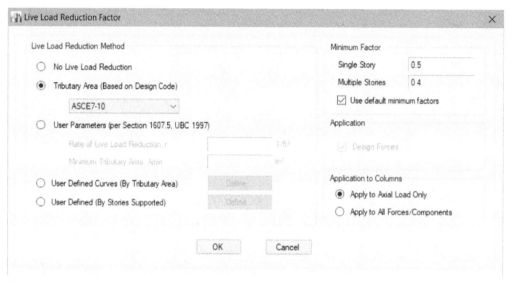

Figure-24. Live Load Reduction Factor dialog box

- Select the **No Live Load Reduction** button if you want no reduction in the live loads.
- Select the **Tributary Area (Based on Design Code)** radio button if you want to set the reduction of load based on the design code selected in the next drop-down.
- Select the **User Parameters (per Section 1607.5, UBC 1997)** radio button if you want to specify the rate of reduction and minimum tributary area manually in the edit boxes below this radio button. Tributary area is a loaded area that contributes to the load on the structural members supporting that area. It can also be called the load periphery.
- Select the **User Defined Curves (By Tributary Area)** radio button to define the curves for load reduction. Click on the Define button, the **Live Load Reduction Factor Curves** dialog box will be displayed; refer to Figure-25. Specify the desired curve data in the **Trib. Area, ft2** and **Red. Factor** edit boxes, and then click on the **Add** button on the right of the table to add it to the curve. Repeat this procedure to create the required curve. Click on the **Add Curve** button to add the next curve if required. Preview of the curve will be displayed in the graph on the right in the dialog box. Click on the **OK** button from the dialog box to apply the changes.

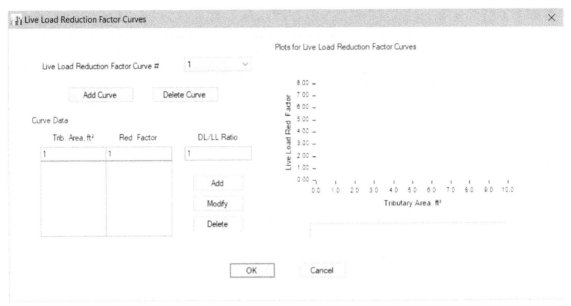

Figure-25. Live Load Reduction Factor Curves dialog box

- Select the **User Defined (By Stories Supported)** radio button if you want to define reduction factor story-wise. Click on the **Define** button, the **Live Load Reduction By Stories Supported** dialog box will be displayed; refer to Figure-26. Specify the desired number of stories supported and reduction factor in respective edit box. Click on the **Add** button to add it in the table. Repeat this procedure to add required number of parameters and then click on the **OK** button.

Figure-26. Live Load Reduction By Stories Supported dialog box

- Specify the desired parameters in the **Minimum Factor** area of the dialog box if you have selected **Tributary Area (Based on Design Code)** or **User Parameters** radio button in the **Live Load Reduction Factor** dialog box.
- If you want to apply the load reduction only to axial forces then select the **Apply to Axial Load only** radio button otherwise select the **Apply to All Forces/Components** radio button.
- Click on the **OK** button from the dialog box to apply the changes.

SETTING LATERAL DISPLACEMENT TARGETS

The **Set Lateral Displacement Targets** tool is used to define the target structural members which would increase in size after application of load. By default, ETABS predicts which members should be increased in size to control the displacements based on the energy per unit volume in the members. The members with more energy per unit volume are increased in size a larger percentage than those with smaller energies per unit volume. Some members with small energy per unit volume may be decreased in size if they are still acceptable for strength considerations. The procedure to use this tool is give next.

- Click on the **Set Lateral Displacement Targets** tool from the **Design** menu. The **Lateral Displacement Targets** dialog box will be displayed; refer to Figure-27.

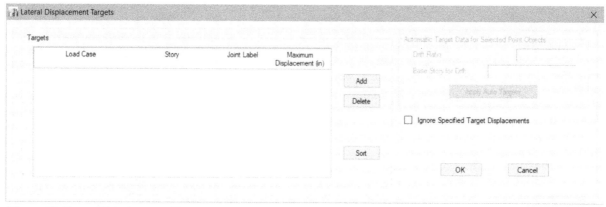

Figure-27. Lateral Displacement Targets dialog box

- Click on the **Add** button to add a new target. A new load case at default joint label will be added in the table. Set the desired parameters in the table. Repeat this step to specify as many targets as you want.
- Select the **Ignore Specified Target Displacements** check box if you want the earlier specified displacement targets by design iterations to be ignored.
- The options in the **Automatic Target Data for Selected Point Objects** area are active only when you have selected joints before selecting this tool. Specify the desired value for drift ratio in the **Drift Ratio** edit box. Select the desired story level from the **Base Story for Drift** drop-down.
- Select the **Apply Auto Parts** button if you want to add auto calculated targets to selected points.
- After setting the desired parameters, click on the **OK** button.

SETTING TIME PERIOD TARGETS

The **Set Time Period Targets** tool is used to set the mode and time period of load. On the basis of time period, the design iteration will modify the structure members. The procedure to use this tool is given next.

- Click on the **Set Time Period Targets** tool from the **Design** menu. The **Time Period Targets** dialog box will be displayed; refer to Figure-28.

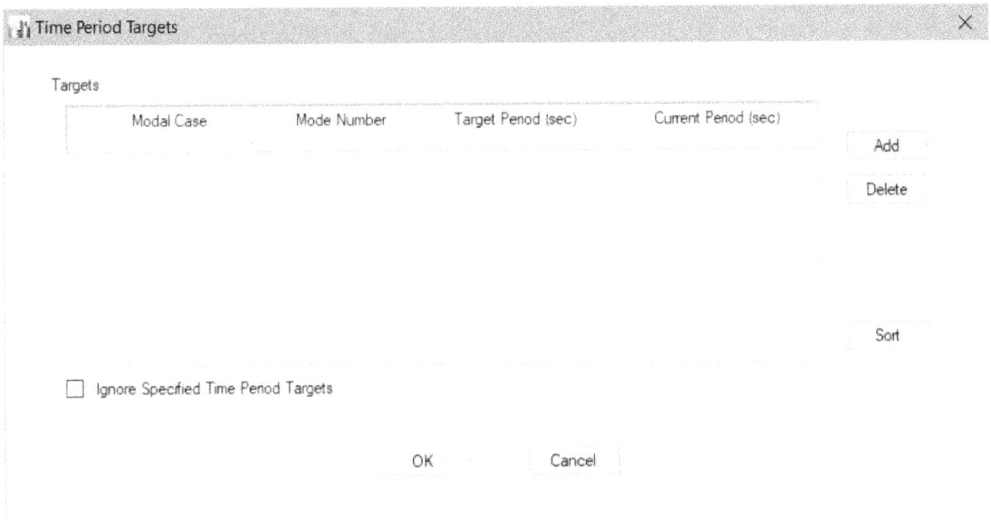

Figure-28. Time Period Targets dialog box

- Click on the **Add** button from the dialog box. A new load case will be added. Specify the parameters as required. Repeat this step to get as many load cases as required.
- Select the **Ignore Specified Time Period Targets** check box if you want to ignore the automatically specified targets during design iteration.
- Click on the **OK** button from the dialog box to apply the changes.

FOR STUDENT NOTES

Chapter 7

Detailing

Topics Covered

The major topics covered in this chapter are:

- *Introduction*
- *Detailing Preferences*
- *Concrete Detailing Preferences*
- *Steel Detailing Preferences*
- *Setting Rebar Selection Rule*
- *Adding/Modifying Slab Sections*
- *Drawing Sheet Setup*
- *Starting Detailing*
- *Show Detailing*
- *Clear Detailing*
- *Exporting Drawing*
- *Printing Drawings*

INTRODUCTION

The tools in the **Detailing** menu are used to create and detail the model; refer to Figure-1. Various tools in this menu are discussed next.

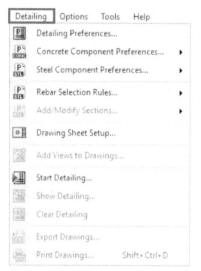

Figure-1. Detailing menu

DETAILING PREFERENCES

The **Detailing Preferences** tool in the **Detailing** menu is used to set preference for detailing of the design. The procedure to use this tool is given next.

- Click on the **Detailing Preferences** tool from the **Detailing** menu. The **Detailing Preferences** dialog box will be displayed; refer to Figure-2.

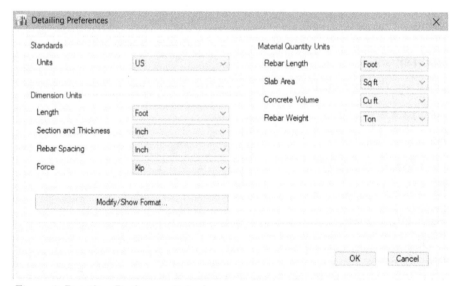

Figure-2. Detailing Preferences dialog box

- Select the desired options from the drop-downs in this dialog box to set the units of detailing.
- After setting desired parameters, click on the **OK** button.

CONCRETE DETAILING PREFERENCES

The Concrete Detailing Preferences are used to define the look and feel of concrete components in the model. The procedure to define preferences is given next.

- Click on the **Slab** tool from the **Concrete Component Preferences** cascading menu of the **Detailing** menu. The **Concrete Detailing Preferences** dialog box will be displayed with **Slabs** tab selected; refer to Figure-3.

Figure-3. Concrete Detailing Preferences dialog box

- Specify the desired options in this tab to define how concrete slabs will be displayed in detailing.
- Set the parameters in other tabs in the same way and click on the **OK** button.

STEEL DETAILING PREFERENCES

The Steel Detailing Preferences are used to define how steel components will be displayed in the model. The procedure is discussed next.

- Click on the **Floor Framing** tool from the **Steel Component Preferences** cascading menu of the **Detailing** menu. The **Steel Detailing Preferences** dialog box will be displayed; refer to Figure-4.
- Set the desired parameter in the dialog box to define how different steel frame components will be created in detailing. After setting the parameters, click on the **OK** button.

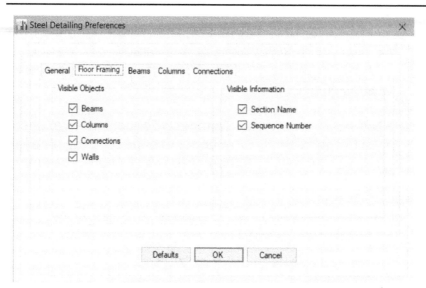

Figure-4. Steel Detailing Preferences dialog box

SETTING REBAR SELECTION RULE

The tools in the **Rebar Selection Rule** cascading menu are used to specify rebar size (max, min, and preferred), rebar quantity (max, min), rebar spacing (max, min), rebar length (for columns only), and rebar installation arrangement (for piers only). The procedure to use tools in this cascading menu are given next.

Setting Rebar Rules for Beams

• Click on the **Beams** tool from the **Rebar Selection Rule** cascading menu. The **Rebar Selection Rules** dialog box will be displayed; refer to Figure-5.

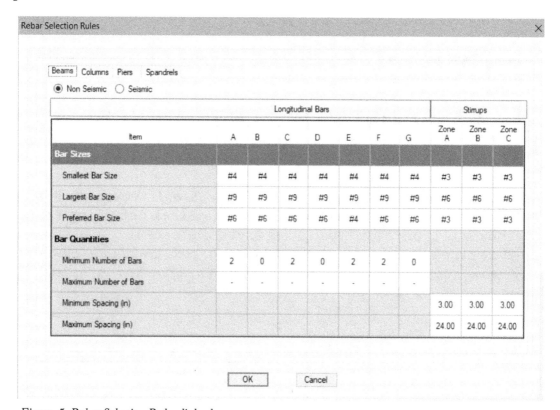

Item			Longitudinal Bars					Stirrups		
	A	B	C	D	E	F	G	Zone A	Zone B	Zone C
Bar Sizes										
Smallest Bar Size	#4	#4	#4	#4	#4	#4	#4	#3	#3	#3
Largest Bar Size	#9	#9	#9	#9	#9	#9	#9	#6	#6	#6
Preferred Bar Size	#6	#6	#6	#6	#4	#6	#6	#3	#3	#3
Bar Quantities										
Minimum Number of Bars	2	0	2	0	2	2	0			
Maximum Number of Bars	-	-	-	-	-	-	-			
Minimum Spacing (in)								3.00	3.00	3.00
Maximum Spacing (in)								24.00	24.00	24.00

Figure-5. Rebar Selection Rules dialog box

- The **Non Seismic** radio button is selected by default. Select the **Seismic** radio button to include bars that are capable of handling seismic load.
- Select the desired bar sizes from the table for various categories like columns, piers, and spandrels.
- After setting desired parameters, click on the **OK** button.

ADDING/MODIFYING SLAB SECTIONS

The **Add/Modify Sections** cascading menu is active only when you have performed detailing by using the **Start Detailing** tool. There is only one tool available in this cascading menu which is **Slab** tool. The procedure to use this tool is given next.

- Click on the **Slab** tool from the **Add/Modify Sections** cascading menu of the **Detailing** menu. The **Add/Modify Slab Sections** window will be displayed; refer to Figure-6.

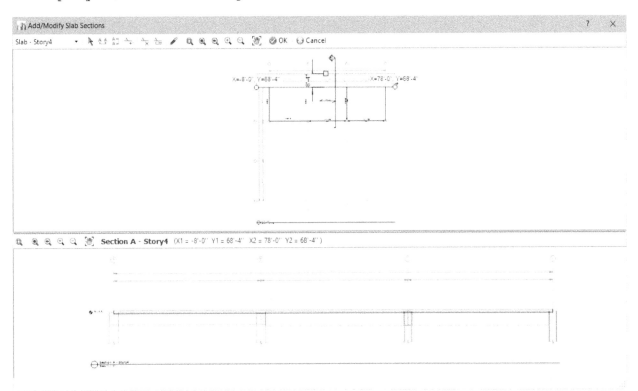

Figure-6. Add/Modify Slab Sections window

- Select the desired slab from the slab drop-down at the top left in the window. The respective slab details will be displayed in the window. The slabs are displayed with their story number.
- Click on the **Add Default Section Along X** button to section the slab along X axis and click on the **Add Default Section Along Y** button to section the slab along Y axis.

Drawing Section Line

- If you want to section the slab at desired location then click on the **Draw Section Line** button from the window. You will be asked to specify the start point of the section line. Click at the desired

location in top window to set starting point and then click at
desired location to specify end point of section line; refer to
Figure-7. The section view will be displayed in the bottom half of
the window; refer to Figure-8.

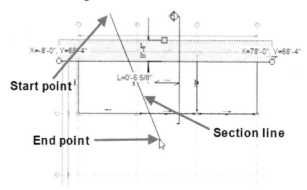

Figure-7. Creating section line

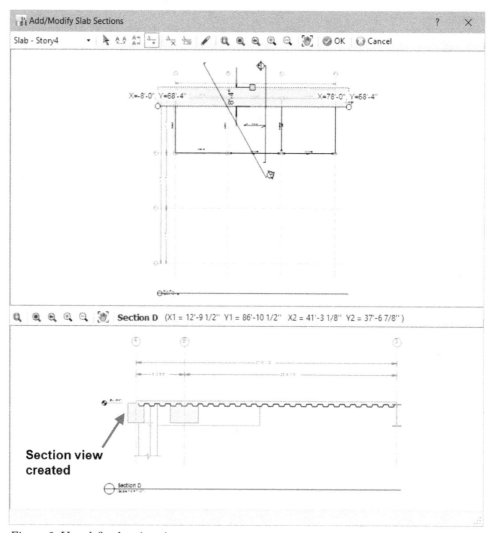

Figure-8. User defined section view

Deleting Selected Sections

- Select the desired section line to be deleted and then click on the
  Delete Selected Section button from the top toolbar of the window.
  The section will be deleted.

Show/Modify Section Line Properties

- Select the desired section line and click on the **Show/Modify Section Line Properties** button from the top of the window. The **Section Line Properties** dialog box will be displayed; refer to Figure-9.

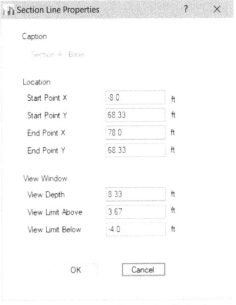

Figure-9. Section Line Properties dialog box

- Set the desired locations of start and end points of section line and set the depth of section in the edit boxes of this dialog box. Click on the **OK** button to apply changes.

The other buttons are used to control view of the sections in the window. Click on the **OK** button from the **Add/Modify Slab Sections** window to set the slab sections in detailing.

DRAWING SHEET SETUP

The **Drawing Sheet Setup** tool in the **Detailing** menu is used to set the size and other parameters of sheet. The procedure to use this tool is given next.

- Click on the **Drawing Sheet Setup** tool from the **Detailing** menu. The **Drawing Sheet Setup** dialog box will be displayed; refer to Figure-10.
- Set the desired sheet type and size from the **Drawing Units and Size** area of the dialog box.
- Set the desired scale from the **Overall Drawing Scale** area.
- Set the other parameters as required and click on the **OK** button.

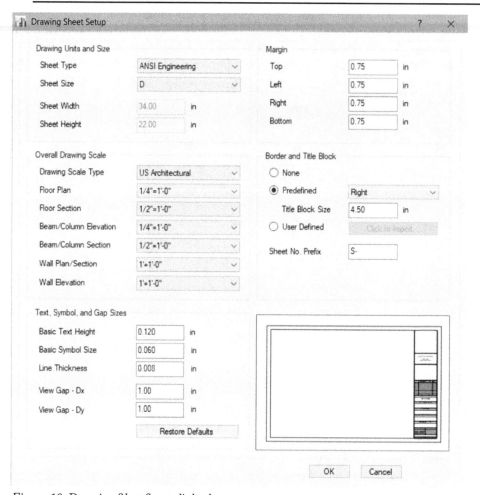

Figure-10. Drawing Sheet Setup dialog box

STARTING DETAILING

The **Start Detailing** tool is used to start detailing various components of structure. The procedure is given next.

- Click on the **Start Detailing** tool from the **Detailing** menu. The process of detailing will start and once it is complete, the framing plan will be displayed as per the preferences set in the **Detailing Preferences** dialog box earlier; refer to Figure-11.

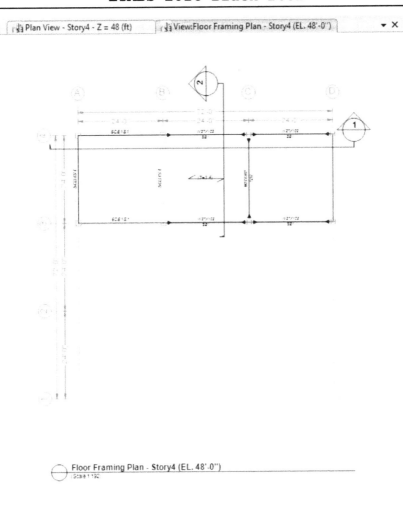

Figure-11. Floor plan created by detailing

Show Detailing

The **Show Detailing** tool in the **Detailing** menu is used to display list of all drawings created by **Start Detailing** tool in the **Detailing** tab of the **Model Explorer**. The **Detailing** tab in **Model Explorer** will be displayed as shown in Figure-12.

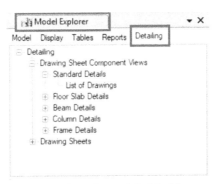

*Figure-12. Detailing tab in Model
Explorer*

Expand the desired category in the tab and double-click to check the drawing; refer to Figure-13.

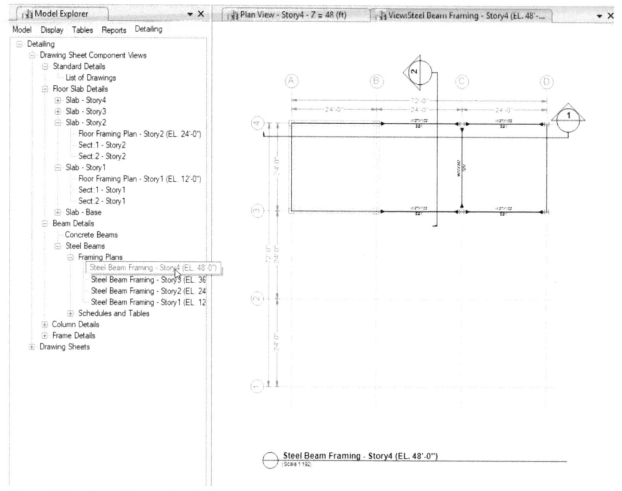

Figure-13. Steel Beam Framing drawing

CLEAR DETAILING

The **Clear Detailing** tool is used to clear all drawing earlier created by **Start Detailing** tool.

EXPORTING DRAWING

The **Export Drawing** tool is used to export drawings created by detailing in DXF or DWG format. The procedure to use this tool is given next.

- After performing detailing, click on the **Export Drawing** tool from the **Detailing** menu. The **Export Drawings** dialog box will be displayed; refer to Figure-14.
- Select the desired radio button from the **Export As** area to define the format in which the drawings will be exported.
- Select the desired radio button from the **Drawing Sheets to Export** area to specify the drawing sheets to be exported.
- Specify the desired location in the **Target Directory** edit box to save exported drawing files.
- After setting the desired parameters, click on the **Start Export** button. The files will be exported at specified location.
- Click on the **Done** button to exit the dialog box.

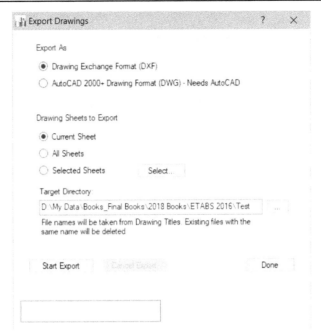

Figure-14. Export Drawings dialog box

PRINTING DRAWINGS

The **Print Drawings** tool in the **Detailing** menu is used to print drawings using the printing devices connected to system. The procedure to use this tool is given next.

- Click on the **Print Drawings** tool from the **Detailing** menu after creating drawings. The **Print Drawings** dialog box will be displayed; refer to Figure-15.

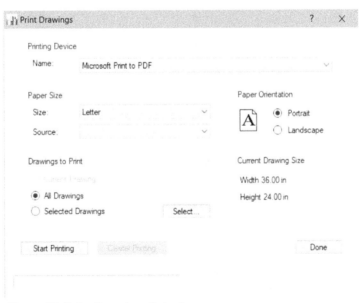

Figure-15. Print Drawings dialog box

- Select the desired printing device from the **Name** edit box.
- Set the paper size, paper orientation and other parameters as required and then click on the **Start Printing** button. The printing will start and once the process is complete, click on the **Done** button.

FOR STUDENT NOTES

Chapter 8

Project

Topics Covered

The major topics covered in this chapter are:

- *Introduction*
- *Project*

INTRODUCTION

In this chapter, we will work on a real world project of structural designing. We will design structure of 5 story building with ground floor used as parking. The material used for structure is concrete. which is

Starting Project

- Start ETABS 2016 from the **Start** menu and click on the **New** button. The **Model Initialization** dialog box will be displayed.
- Select **U.S. Customary** option from the **Display Units** drop-down. Select the **AISC14** option from the **Steel Section Database** drop-down. Select the **AISC 360-10** option from the **Steel Design Code** drop-down. Select the **ACI 318-14** option from the **Concrete Design Code** drop-down.
- After setting desired parameters, click on the **OK** button. The **New Model Quick Templates** dialog box will be displayed; refer to Figure-1.

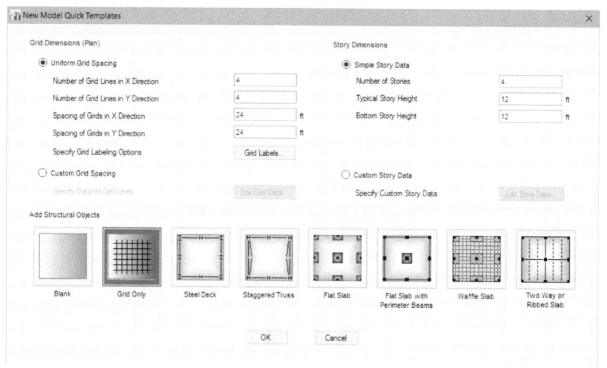

Figure-1. New Model Quick Templates dialog box

- Set the number of stories as **5** in the **Number of Stories** edit box. Set typical story height and bottom story height as **12** in respective edit boxes.
- Keep the other parameters as default and make sure the **Grid Only** button is selected.
- Click on the **OK** button. The grid of building will be displayed; refer to Figure-2.

Figure-2. Grid created for project

Adding Material

- Click on the **Material Properties** tool from the **Define** menu. The **Define Materials** dialog box will be displayed; refer to Figure-3.

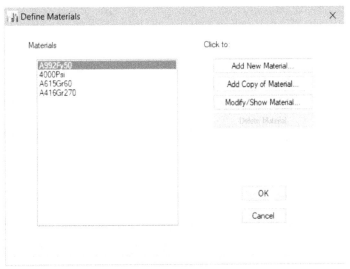

Figure-3. Define Materials dialog box

- Click on the **Add New Material** button from the dialog box. The **Add New Material Property** dialog box will be displayed; refer to Figure-4.

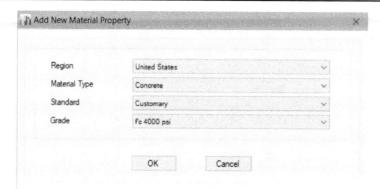

Figure-4. Add New Material Property dialog box

- Select the **United States** option from **Region** drop-down, **Concrete** option from **Material Type** drop-down, **Customary** option from the **Standard** drop-down and **f'c 4000 psi** from the **Grade** drop-down.
- After setting the parameters, click on the **OK** button. The **Material Property Data** dialog box will be displayed; refer to Figure-5.

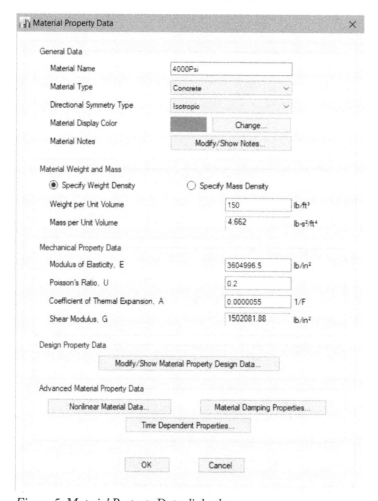

Figure-5. Material Property Data dialog box

- Set the parameters as shown in Figure-5 and click on the **OK** button. The material will be created and displayed in the **Define Materials** dialog box.
- Click on the **OK** button from the **Define Materials** dialog box.

Creating Frame Section

- Click on the **Frame Sections** tool from the **Section Properties** cascading menu of the **Define** menu. The **Frame Properties** dialog box will be displayed.
- Click on the **Add New Property** button from the dialog box. The **Frame Property Shape Type** dialog box will be displayed; refer to Figure-6.

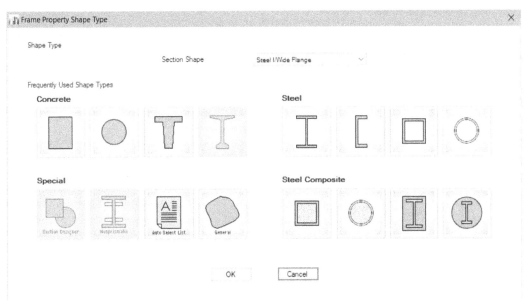

Figure-6. Frame Property Shape Type dialog box

- Select the **Concrete Rectangular** option from the **Section Shape** drop-down at the top in the dialog box. The **Frame Section Property Data** dialog box will be displayed; refer to Figure-7.

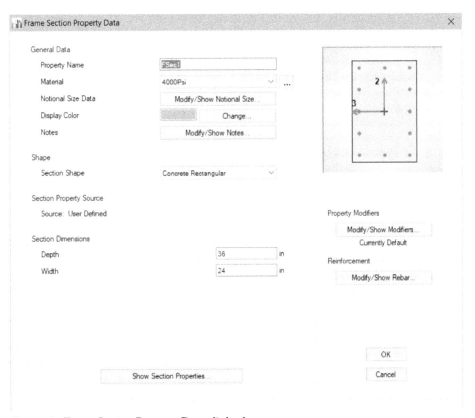

Figure-7. Frame Section Property Data dialog box

- Set the name **Conc Column** for frame property in the **Property Name** edit box.
- Select the newly created material **4000Psi** from the **Material** drop-down.
- Click on the **Modify/Show Rebar** button in the **Reinforcement** area of the dialog box. The **Frame Section Property Reinforcement Data** dialog box will be displayed; refer to Figure-8.

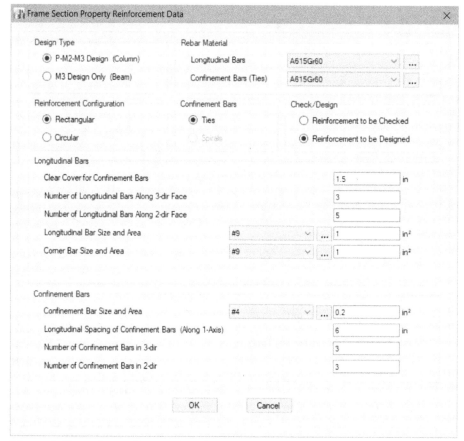

Figure-8. Frame Section Property Reinforcement Data dialog box

- Select the **P-M2-M3 Design (Column)** radio button from the **Design Type** area.
- Select **Circular** radio button from the **Reinforcement Configuration** area.
- Set the other parameters as shown in Figure-8 and click on the **OK** button. The **Frame Section Property Data** dialog box will be displayed again.
- Set the depth of column as **36** inch and width as **24** inch in respective edit boxes of the dialog box. Click on the **OK** button to create the property.
- Create the **Conc Beam** property in the same way with reinforcement options set as shown in Figure-9. Make sure the size is 36 x 24 inches.

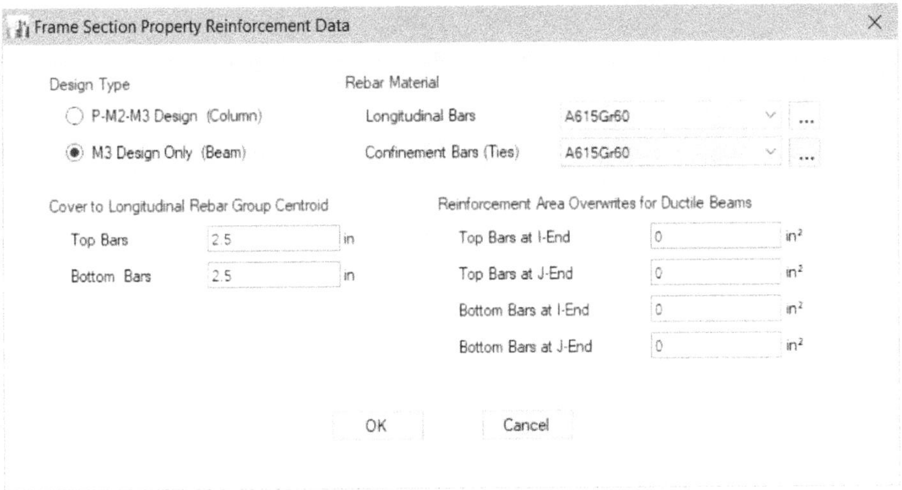

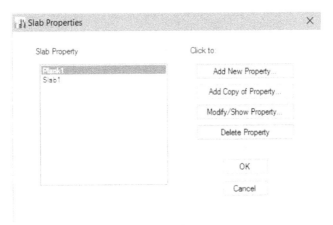

Figure-9. Frame Section Property Reinforcement Data dialog box for beam

- After creating frame section properties, click on the **OK** button.

Defining Slab Section

- Click on the **Slab Sections** tool from the **Section Properties** cascading menu of the **Define** menu. The **Slab Properties** dialog box will be displayed; refer to Figure-10.

Figure-10. Slab Properties dialog box

- Click on the **Add New Property** button. The **Slab Property Data** dialog box will be displayed; refer to Figure-11.

Figure-11. Slab Property Data dialog box

- Specify the name of slab as **Floor slab** in the **Property Name** edit box.
- Set the thickness of slab as **9** inch and other parameters as in Figure-11.
- After setting the parameters, click on the **OK** button. A new property will be added in the slab properties. Click on the **OK** button from the **Slab Properties** dialog box.

Defining Deck Section

- Click on the **Deck Sections** tool from the **Section Properties** cascading menu of the **Define** menu. The **Deck Properties** dialog box will be displayed.
- Click on the **Add New Property** tool from the dialog box. The **Deck Property Data** dialog box will be displayed; refer to Figure-12.
- Set the name of property as **Roof** in the **Property Name** edit box.
- Select the **Filled** option from the **Type** drop-down and set the other parameters as shown in Figure-12.

Figure-12. Deck Property Data dialog box

- After setting desired parameters, click on the **OK** button. The property will be added in the **Deck Properties** dialog box. Click on the **OK** button.

Defining Wall Section

- Click on the **Wall Sections** tool from the **Section Properties** cascading menu of the **Define** menu. The **Wall Properties** dialog box will be displayed.
- Click on the **Add New Property** button. The **Wall Property Data** dialog box will be displayed; refer to Figure-13.
- Set the name of property as **Wall section** in the **Property Name** edit box.

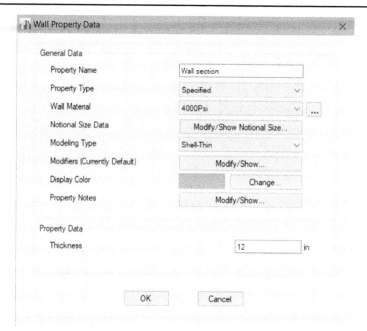

Figure-13. Wall Property Data dialog box

- Click on the **Add Material** button next to **Wall Material** drop-down. The **Define Materials** dialog box will be displayed.
- Click on the **Add New Material** button and set the properties as shown in Figure-14. Click on the **OK** button. The **Material Property Data** dialog box will be displayed.

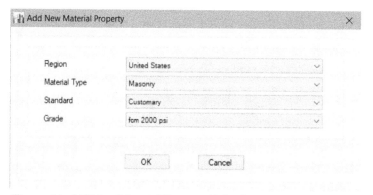

Figure-14. Add New Material Property dialog box

- Specify the name of material as **Wall Material** in the **Material Name** edit box and set the parameters as shown in Figure-15.
- Click on the **OK** button from the dialog box. The **Define Materials** dialog box will be displayed again. Click on the **OK** button. The **Wall Property Data** dialog box will be displayed again.
- Select **Wall Material** option from the **Wall Material** drop-down.
- Specify the thickness of wall as **9** inch in the **Thickness** edit box.
- Click on the **OK** button from the **Wall Property Data** dialog box to create wall section.

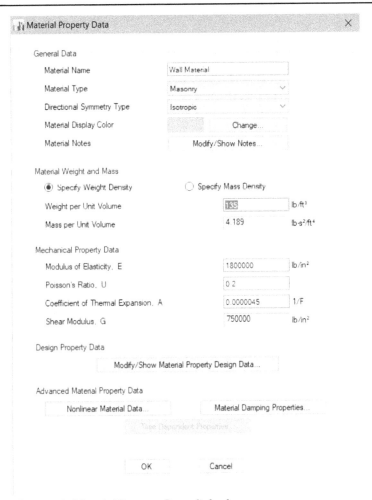

Figure-15. Material Property Data dialog box

- The **Wall Properties** dialog box will be displayed again. Click on the **OK** button from the dialog box to exit.

Creating Structure

- Select the **All Stories** option from the **Stories** drop-down at the bottom right corner of the window.
- Click on the **Quick Draw Beams/Columns (Plan, Elev, 3D)** tool from the **Draw Beam/Column/Brace Objects** cascading menu of the **Draw** menu. The **Properties of Object** tab will be displayed below **Model Explorer**.
- Select the **Conc Column** option from the **Property** drop-down and select all the vertical lines of top floor from the 3D View. The concrete columns will be created; refer to Figure-16.

Figure-16. Concrete columns created

- Select the **Conc Beam** option from the **Property** drop-down and select the lines along X and Y axes in the Plan view. The concrete beams will be created; refer to Figure-17.

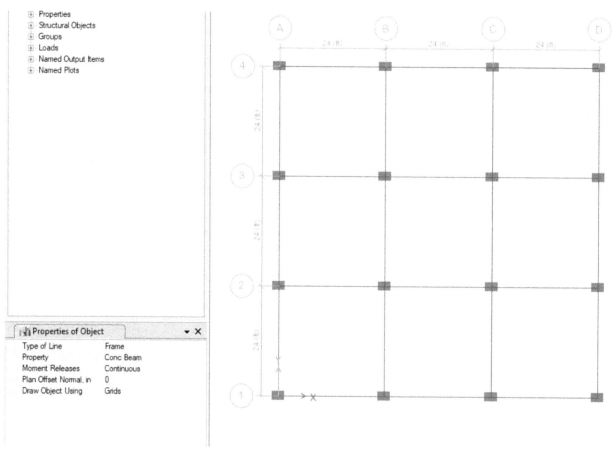

Figure-17. Concrete beams created

- Click on the **Quick Draw Floor/Wall (Plan, Elev)** tool from the **Draw Floor/Wall Objects** cascading menu of the **Draw** menu. The **Properties of Object** tab will be displayed and you will be asked to specify location.
- Select the **Floor slab** option from the **Property** drop-down and click in the rectangular boxes. The model will be displayed as shown in Figure-18.

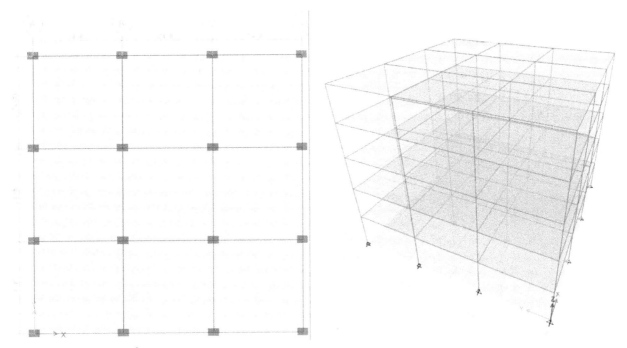

Figure-18. Floors created

- Select the **One Story** option from the **Stories** drop-down and select the floor slabs of top floor created recently from the Plan view; refer to Figure-19.

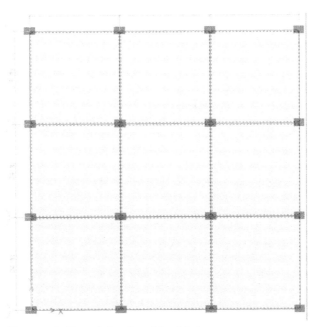

Figure-19. Floor slabs selected for deletion

- Press the **DELETE** button from the keyboard to delete floor slabs created at the top.
- Click on the **Quick Draw Floor/Wall (Plan, Elev)** tool from the **Draw Floor/Wall Objects** cascading menu of the **Draw** menu again. Select the **Roof** property from the **Property** drop-down.
- Click in the rectangular boxes of plan view which you deleted earlier. The roof slabs will be created.
- Click on the **Quick Draw Walls (Plan)** tool from the **Draw Floor/Wall Objects** cascading menu of the **Draw** menu. You will be asked to specify location for creating wall.
- Select **All Stories** option from the **Stories** drop-down at the bottom right corner of the window.
- Select **Wall section** option from the **Property** drop-down and select the outer edges of the building. The walls will be created as shown in Figure-20. Press **ESC** to exit the tool. This completes the structure.

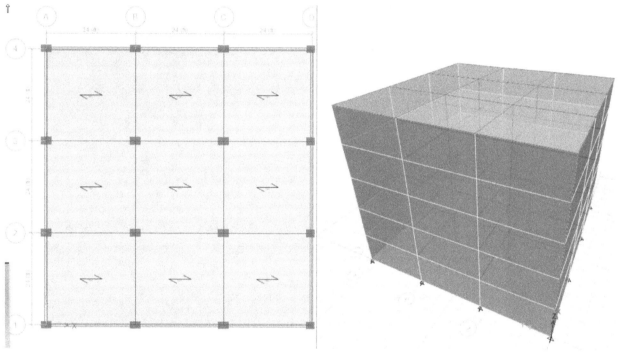

Figure-20. Walls created in project

Assigning Loads

- Click on the **Load Patterns** tool from the **Define** menu. The **Define Load Patterns** dialog box will be displayed.
- Enter the parameters as shown in Figure-21 and click on the **Add New Load** button. Wind load will be added as per ASCE 7-10.

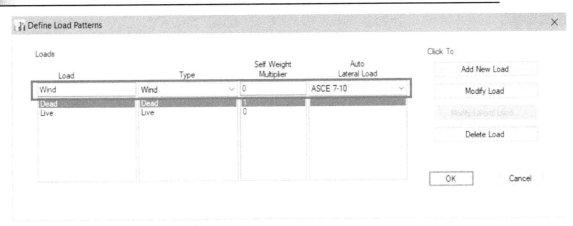

Figure-21. Wind load pattern created

- Click on the **OK** button from the dialog box to create the load pattern.
- Click on the **Uniform** tool from the **Shell Loads** cascading menu of the **Assign** menu. The **Shell Load Assignment - Uniform** dialog box will be displayed.
- Select the **Live** option from the **Load Pattern Name** drop-down. Specify the load value as **25** in **Load** edit box.
- Select **One Story** option from the **Stories** drop-down. Select the top roof shells; refer to Figure-22.

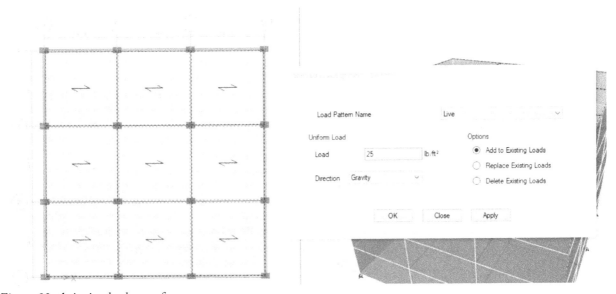

Figure-22. Assigning load on roof

- Make sure you select **Add to Existing Loads** radio button and then click on the **OK** button.
- Click on the **Floor slab** object in the **Slab Sections** node of the **Model** tab in the **Model Explorer** and right-click on it. A shortcut menu will be displayed; refer to Figure-23.

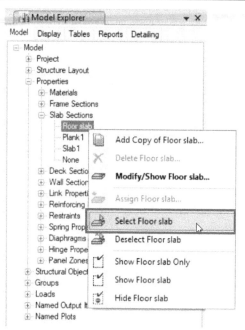

Figure-23. Selecting floor slabs

- Click on the **Select Floor slab** option from the shortcut menu. All the floor slabs will be selected.
- Click on the **Uniform** tool from the **Shell Loads** cascading menu of the **Assign** menu. The **Shell Load Assignment-Uniform** dialog box will be displayed.
- Specify the load value as **100** lb/ft^2 and select the **Add to Existing Loads** radio button.
- Make sure the **Live** option is selected in the **Load Pattern Name** edit box and click on the **OK** button.

Setting Mesh Size

- Click on the **Automatic Mesh Settings for Floors** tool from the **Analyze** menu to define mesh size for floors. The **Automatic Mesh Options (for Floors)** dialog box will be displayed.
- Set the mesh size as **4** feet in the **Approximate Maximum Mesh Size** edit box. Select the **Use Localized Meshing** and **Merge Joints where Possible** check boxes.
- Click on the **OK** button to specify the size of floor mesh.
- Click on the **Automatic Rectangular Mesh Settings for Walls** tool from the **Analyze** menu to define mesh size of walls. The **Automatic Rectangular Mesh Options (for Walls)** dialog box will be displayed.
- Specify the size of mesh in the dialog box and click on the **OK** button.

Running Analysis

- Click on the **Set Load Cases to Run** tool from the **Analyze** menu. The **Set Load Cases to Run** dialog box will be displayed; refer to Figure-24.

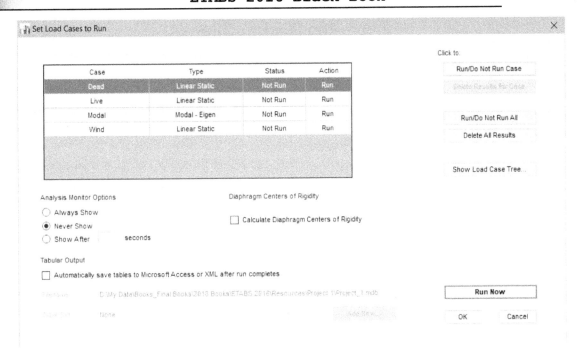

Figure-24. Set Load Cases to Run dialog box

- Make sure all the cases are selected and select the **Always Show** radio button from the **Analysis Monitor Options** area.
- Click on the **Run Now** button. Once the analysis is performed, the deformed model will be displayed in 3-D View; refer to Figure-25.

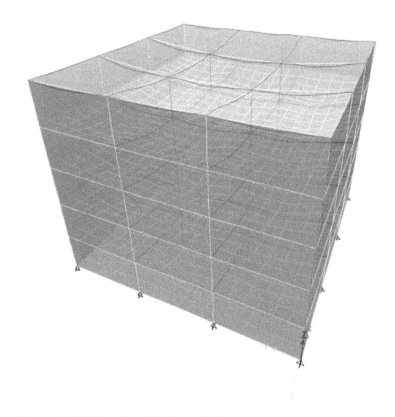

Figure-25. Deformed model after analysis

Performing Design Check

- Click on the **View/Revise Preferences** tool from the **Concrete Frame Design** cascading menu of the **Design** menu. The **Concrete Frame Design Preferences** dialog box will be displayed.

- Select the **ACI 318-14** option from the **Design Code** field of the **Concrete Frame Design** dialog box and click on the **OK** button.

- Click on the **Start Design/Check** tool from the **Concrete Frame Design** cascading menu of **Design** menu. The size of different reinforcements of concrete beams will be displayed in the plan view; refer to Figure-26.

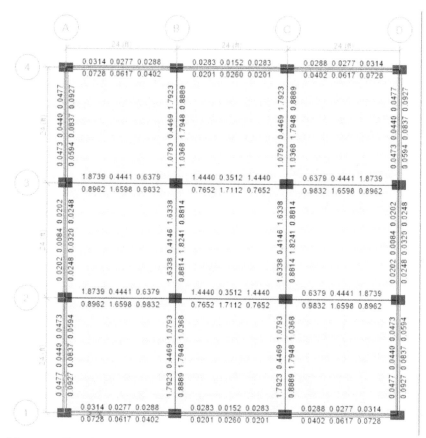

Figure-26. Reinforcement of concrete beam

- Click on the **Verify All Members Passed** tool from the **Concrete Frame Design** cascading menu of the **Design** menu to make sure all the concrete members have passed design check. The message box will be displayed as shown in Figure-27. Click on the **OK** button.

Figure-27. ETABS message box

Detailing

- Click on the **Detailing Preferences** tool from the **Detailing** menu. The **Detailing Preferences** dialog box will be displayed.
- Set the unit system to **US** from the **Units** drop-down and click on the **OK** button to apply preferences.
- Click on the **Start Detailing** tool from the **Detailing** menu. The **Detailing** tab in **Model Explorer** will become active and drawing will be displayed in Floor Framing Plan; refer to Figure-28.

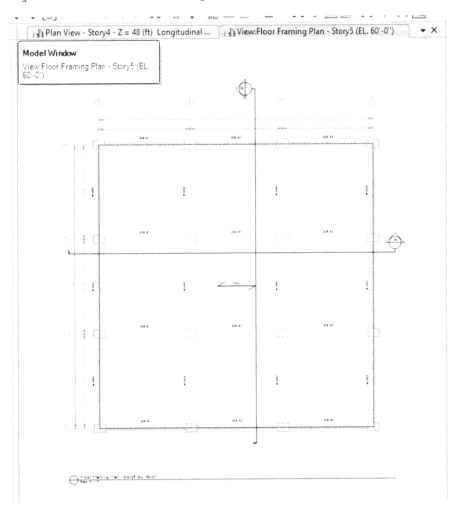

Figure-28. Floor Framing Plan

- Select the **Detailing** node from the **Detailing** tab of **Model Explorer** and click on the **Export Drawings** tool from **Detailing** menu. The **Export Drawings** dialog box will be displayed; refer to Figure-29.

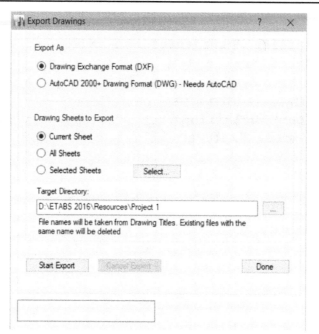

Figure-29. Export Drawings dialog box

- Set the DXF format and set the desired location.
- Click on the **Start Export** button. Once the drawings are exported, click on the **Done** button.

Save the file and close the project by using the **Close** button.

Index

A

Add Copy of Property button 3-6
Add Default Design Combos button 5-25
Add Grid Lines at Selected Joints tool 2-4
Add Image button 1-28
Additional Mass tool 4-5, 4-17, 4-31
Add/Modify Sections cascading menu 7-5
Add New Combo button 5-24
Add New Load button 5-2
Add New Section Property button 3-5
Add New User Report tool 1-31
Add Note button 1-28
Add Notional load cases into seismic combos drop-down 6-4
Add Soil Layer button 3-35
Add Structural Objects area 1-8
Add XML Property Files dialog box 3-4
Advanced Link P-Delta Parameters dialog box 3-28
Advanced SAPFire Options tool 5-32
Analysis Method drop-down 6-4
Analysis Model for Nonlinear Hinges tool 5-34
Architectural Plan Import dialog box 1-24
Area Spring tool 4-30
Auto Construction Sequence Load Case tool 5-26
Auto Draw Cladding tool 2-22
Automatic Mesh Settings for Floors tool 5-33
Automatic Rectangular Mesh Settings for Walls tool 5-34

B

Beams tool 7-4
Buckling load case 5-22
Buckling option 5-8

C

Capture Picture cascading menu 1-32
Change Design Section tool 6-12
Check Model tool 5-29
Clear Detailing tool 7-10
Client Name edit box 1-14
Close tool 1-20
Column/Brace Rebar Ratio for Creep Analysis tool 4-22
Column Splice Overwrite tool 4-20
Complete Quadratic Combination option 5-17
Concrete Design Code 1-6
Concrete Frame Design cascading menu 6-14
Continue from State at End of Nonlinear Case (Loads at End of Case ARE Included) radio button 5-12
Convert Combos to Nonlinear Cases button 5-26
Convert to SD Section button 3-8
Copy Assigns tool 4-44
Covert to User Combinations (Editable) check box 5-26
CQC3 5-17

Cracking Analysis Options tool 5-35
Create Report cascading menu 1-30
Create Video cascading menu 1-26
Cumulative Energy Components tool 5-44
Custom Grid Spacing radio button 1-7
Custom Story Data radio button 1-8

D

Dead Load Pattern drop-down 1-10
Deck/Floor drop-down 1-10
Deck Sections tool 3-25
Deck Section tool 4-24
Deformed Shape tool 5-39
Delete Design Results tool 6-14
Delete Multiple Properties button 3-7
Delete Property button 3-7
Delete Selected Section button 7-6
Design Preferences node 1-14
Detailing Preferences tool 7-2
Detailing tab 1-19
Diaphragms tool 3-37, 4-4, 4-28
Direct Analysis Method 6-4
Display Design Info tool 6-11
Display menu 5-38
Display Performance Check tool 5-43
Display tab 1-16
Display Units 1-6
Distributed tool 4-37
Double Sum method 5-17
Draw Beam/Column/Brace Objects cascading menu 2-7
Draw Beams/Columns/Braces (Plan 2-8
Draw Channel tool 3-16
Draw Defined Section tool 3-16
Draw Design Strips tool 2-15
Draw Developed Elevation Definition tool 2-19
Draw Dimension Lines tool 2-16
Draw Floor/Wall Objects cascading menu 2-11
Draw Floor/Wall (Plan 2-12
Draw Grids tool 2-5
Drawing Sheet Setup tool 7-7
Draw Joint Objects tool 2-3
Draw Links tool 2-15
Draw Rectangular Floor/Wall (Plan 2-12
Draw Reference Circle tool 3-14
Draw Reference Line tool 3-14
Draw Reference Planes tool 2-19
Draw Reference Points tool 2-17
Draw Reference Point tool 3-13
Draw Section Line button 7-5
Draw Tendons tool 2-15
Draw Using Snap Only tool 2-7
Draw Walls (Plan) tool 2-14
Draw Wall Stacks (Plan 2-21

E

Edge Releases tool 4-29
Edit Grid Data button 1-7
Edit Stories and Grid System option 2-3

Edit Story and Grid Systems button 1-15
Edit Story Data button 1-8
Effective Length method 6-4
Eigen option 5-7
End Length Offset tool 4-10
Energy/Virtual Work Diagram tool 5-43
Evaluation Versions link 1-3
Export cascading menu 1-25
Export Drawing tool 7-10
Export ETABS/Revit Structure Exchange File dialog box
        1-25
Export to Word Document button 1-30
Export to XML File tool 3-10
Exposure from Extents of Diaphragms radio button 5-3
Exposure from Frame and Shell Objects radio button 5-4

F

File menu 1-20
Floor Diaphragm Rigidity 1-10
Floor Framing tool 7-3
Force/Stress Diagrams cascading menu 5-39
Frame Assignment - Moment Frame Beam Connection
        Type dialog box 1-9
Frame Auto Mesh Options tool 4-18
Frame/Pier/Spandrel/Link Forces tool 5-41
Frame Properties dialog box 3-2
Frame Property Shape Type dialog box 3-3
Frame Section tool 3-2
Frame/Wall Nonlinear Hinges tool 3-29
Framing Type drop-down 6-3

G

General Modal Combination option 5-17
Geometric Nonlinearity Option drop-down 5-12
Glue Joints to Grids option 2-3
Gravity Beam drop-down 1-10
Gravity Column drop-down 1-10
Grid Labels button 1-6
Grid Options cascading menu 2-2
Grids Only button 1-8
Grid System Data dialog box 1-7
Grid System drop-down 1-12
Ground Displacement tool 4-35
Group Definitions tool 3-38
GUI process 5-33
Gust Factor edit box 5-5

H

Hinge Overwrites tool 4-16
Hinges tool 4-15
Hyperstatic Load Case 5-23
Hyperstatic option 5-8

I

Import cascading menu 1-21
Import New Properties button 3-3
Include Effects of Joint Temperatures check box 4-38

Include Rigid Response check box 5-17
Include Selected Joint Objects In Mesh check box 4-7
Insertion Point tool 4-11, 4-27
Interaction Surface tool 3-19
Interactive Design tool 6-11
Isolated Column Footings tool 3-36
Iterative - Based on Loads radio button 5-10

J

Joint Assignment-Additional Mass dialog box 4-5
Joint Assignment - Diaphragms dialog box 4-4
Joint Assignment - Spring dialog box 4-4
Joint cascading menu 4-2
Joint Floor Meshing Options tool 4-6
Joint Load Assignment-Force dialog box 4-34
Joint Load Assignment- Temperature dialog box 4-35
Joint Loads cascading menu 4-33
Joint Story Assignment check box 5-29

L

Last Analysis Run Log tool 5-38
Lateral Beam drop-down 1-10
Lateral Bracing tool 6-6
Lateral Column drop-down 1-10
Linear Static option 5-8
Line Drawing Type drop-down 2-8
Line Spring tool 4-17
Link Properties tool 4-32
Link/Support Properties tool 3-27
Live Load Pattern drop-down 1-10
Live Load Reduction Factors tool 6-18
Load Application field 5-12
Load area 1-10
Load Assigns cascading menu 5-39
Load Cases tool 5-8
Load Combinations 5-24
Load Pattern tool 5-2
Local Axes tool 4-12, 4-30
Lock Onscreen Grid System Edit option 2-3

M

Make Auto Select Section Null tool 6-12
Material Overwrites tool 4-21
Modal Cases tool 5-6
Modal Damping Ratio edit box 5-29
Model Alive tool 5-36
Model Explorer 1-12
Model Initialization dialog bo 1-5
Model tab 1-12
Modify Undeformed Geometry tool 5-37
Moment-Curvature Curve tool 3-20
Moment Frame Beam Connection Type tool 4-19
Moment Frame Type area 1-9
Multiple States radio button 5-13
Multi-Response Case Design field 6-2

N

New button 1-5
New Model Quick Templates dialog box 1-6
Non-iterative-Based on Mass radio button 5-10
Nonlinear Staged Construction option 5-8
Nonlinear Static option 5-8
Nonprismatic Property Parameters tool 4-21
Non Seismic radio button 7-5
Non-uniform tool 4-41
NRC 10 Pct 5-17

O

Opening tool 4-25
Open Structure Wind Parameters tool 4-38
Output Stations tool 4-14
Overwrite Frame Design Procedure tool 6-15

P

Page Setup button 1-29
Panel Zone tool 3-31, 4-5
Parapet Height edit box 5-6
Paste Assigns tool 4-44
P-Delta plus Large Displacements option 5-12, 5-27
Peak Acceleration Threshold (Percentage of Gravity) area 5-29
Peak Load Factor edit box 5-29
Pier Labels tool 3-37
Pier Label tool 4-18
Plot Functions tool 5-45
Point Spring Properties dialog box 4-4
Point Springs tool 3-33
Point tool 4-36
Print Drawings tool 7-11
Printer Setup button 1-29
Print Graphics tool 1-27
Project Information dialog box 1-13
Project Information tool 1-32
Properties of Object form 2-22
Property Modifiers tool 4-8

Q

Quasi-Static radio button 5-13
Quick Draw Beams/Columns (Plan 2-10
Quick Draw Braces (Plan 2-11
Quick Draw Floor/Wall (Plan 2-13
Quick Draw Secondary Beams (Plan 2-10

R

Rebar Selection Rule cascading menu 7-4
Reference Line cascading menu 3-13
Regenerate Report button 1-30
Reinforcing Bar Sizes tool 3-26
Releases/Partially Fixity tool 4-9
Reports tab 1-19
Reset All Overwrites tool 6-14
Reset Design Section to Last Analysis tool 6-13

Reshape Mode option 3-17
Response Spectrum option 5-8
Restraints at Bottom 1-10
Restraint tool 4-2
Results Saved for Nonlinear Static Case dialog box 5-13
Revit Structure .exr File option 1-23
Ritz Modal Case SubType 5-7
Ritz Modes per Step edit box 5-29
Run Analysis tool 5-36

S

Save As tool 1-21
Save tool 1-21
Secondary Beam drop-down 1-10
Secondary Beams area 1-8
Section Designer button 3-9
Section Designer window 3-11
Section Properties cascading menu 3-2
Section Properties tool 3-18
Seismic Design Category drop-down 6-3
Seismic radio button 7-5
Select Design Combinations tool 6-7
Select Design Group tool 6-7
Self-weight Multiplier edit box 5-2
Set Active Degrees of Freedom tool 5-30
Set Default 3D View tool 2-17
Set Elevation View tool 2-18
Set Lateral Displacement Targets tool 6-20
Set Load Cases To Run tool 5-31
Set Plan View tool 2-17
Set Time Period Targets tool 6-20
Shear Wall Design cascading menu 6-17
Shell Assignment-Insertion Point dialog box 4-27
Shell cascading menu 4-22
Shell Loads cascading menu 4-39
Shell Stresses/Forces tool 5-42
Show Detailing tool 7-9
Show/Modify Section Line Properties button 7-7
Show Project Report tool 1-30
Show Table button 1-18
Simple Story Data radio button 1-7
Slab Property Layer Definition Data dialog box 3-24
Slab Sections tool 3-22
Slab Section tool 4-23
Slab tool 7-3
Snap Options tool 2-6
Snap to Grid Intersections & Points button 2-7
Soil Pressure tool 5-40
Soil Profile tool 3-34
Spring Properties cascading menu 3-33
Spring tool 4-3
Square Root of Sum of Squares option 5-17
Start Design/Check tool 6-8
Start Detailing tool 7-8
Steel Deck button 1-8
Steel Design Code 1-6

Steel Frame Design cascading menu 6-2
Steel Frame Design Preferences dialog box 6-2
Steel Frame Preferences 6-2
Steel Section Database 1-6
Steel Stress Check Information dialog box 6-9
Stiffness Modifiers tool 4-25
Stiffness Reduction Method 6-4
Story Data dialog box 1-8
Story Dimensions 1-7
Story drop-down 1-12
Story Response Plots tool 5-45
Structure Layout 1-15
Support/Spring Reactions tool 5-39

T

Tables tab 1-18
Temperature tool 4-35
Tendon Losses tool 4-43
Tendon Property Assign dialog box 4-33
Tendon Sections tool 3-21
Tension/Compression Limits tool 4-14
Thickness Overwrites tool 4-26
Time History load case 5-17
Time History option 5-8
Time Integration methods 5-21
Topographical Factor 5-5

U

Undeformed Shape tool 5-38
Uniform Grid Spacing radio button 1-6
Uniform Load Sets tool 4-39
Uniform Shell Load tool 4-41
Units button 1-12
Unlock Model button 5-38
Upload to CSI Cloud option 1-26
Use Built-in Settings with radio button 1-5
Use Preset P-Delta Settings radio button 5-9
User Defined (By Stories Supported) radio button 6-19
User Defined Curves (By Tributary Area) radio button 6-18
Use Saved User Default Settings option 1-5

V

Verify All Members Pass tool 6-14
Verify Analysis vs Design Section tool 6-13
View/Revise Overwrite tool 6-5
View/Revise Preferences tool 6-2

W

Walking Frequency (Steps/sec) edit box 5-29
Walking Vibrations tool 5-28
Wall Sections tool 3-26
Wall Section tool 4-24
Weight of Person Walking edit box 5-29
Wind Coefficients area 5-5
Wind Pressure Coefficients area 5-4
Working plane drop-down 1-12

Z

Zero Initial Conditions - Start from Unstressed State radio button 5-11

OTHER BOOKS BY CADCAMCAE WORKS

Autodesk Inventor 2018 Black Book

Autodesk Fusion 360 Black Book

AutoCAD Electrical 2015 Black Book
AutoCAD Electrical 2016 Black Book
AutoCAD Electrical 2017 Black Book
AutoCAD Electrical 2018 Black Book

SolidWorks 2014 Black Book
SolidWorks 2015 Black Book
SolidWorks 2016 Black Book
SolidWorks 2017 Black Book
SolidWorks 2018 Black Book

SolidWorks Simulation 2015 Black Book
SolidWorks Simulation 2016 Black Book
SolidWorks Simulation 2017 Black Book
SolidWorks Simulation 2018 Black Book

SolidWorks Flow Simulation 2018 Black Book

SolidWorks Electrical 2015 Black Book
SolidWorks Electrical 2016 Black Book
SolidWorks Electrical 2017 Black Book
SolidWorks Electrical 2018 Black Book

Mastercam X7 for SolidWorks 2014 Black Book
Mastercam 2017 for SolidWorks Black Book

Creo Parametric 3.0 Black Book
Creo Parametric 4.0 Black Book

Creo Manufacturing 4.0 Black Book

ETABS 2016 Black Book

www.ingramcontent.com/pod-product-compliance
Lightning Source LLC
LaVergne TN
LVHW060138070326
832902LV00018B/2839